Collusion, Local Governments and Development in China

Huihua Nie

Collusion, Local Governments and Development in China

A Reflection on the China Model

palgrave
macmillan

Huihua Nie
Renmin University of China
Beijing, China

Translated by Haifeng Li and Ping Chen

ISBN 978-981-10-5058-9 ISBN 978-981-10-5059-6 (eBook)
DOI 10.1007/978-981-10-5059-6

Library of Congress Control Number: 2017945790

Cover credit: TAO Images Limited/Alamy Stock Photo

Printed on acid-free paper

This Palgrave Macmillan imprint is published by Springer Nature
The registered company is Springer Nature Singapore Pte Ltd.
The registered company address is: 152 Beach Road, #21-01/04 Gateway East, Singapore 189721, Singapore

To Those Concerned About Destiny of China
and Promoting China's Progress

Acknowledgements

I avail myself of this opportunity to express my heartfelt gratitude for the invaluable support that I have received over the past 3 years (2009–2012), during which I spent every day scrambling up the mountains of my academic research. I must first mention the China Scholarship Council (CSC) which gave me financial support for my one-year postdoctoral research at Harvard University with Oliver Hart, who, as an internationally luminary economist in contract theory and the theory of the firm, benefited me a lot both in my academic research and teaching.

A number of people have been wonderfully helpful in providing material and research approach for the book. I would very much like to thank Ruixue Jia, Minjie Jiang, Ting Jiang, Jinbo Li, Xianghong Wang, Rudai Yang, Yu Zhang, Yuxiao Zhang, and Zhaoyun Zou, with whom I collaborated well and successively published a series of papers on the topic of collusion and the China model in *Economic and Political Studies*, *Journal of Comparative Economics*, *Economic Research Journal*, *The Journal of World Economy*, *Review of Economics and Statistics*, etc. Many of the papers inspired and enlightened me throughout the journey of completing this book.

I also give my thanks to my students Li Dang, Zhibo Tang and Hua Guo for their arduous and down-to-earth assistant research for me. My thanks would also go to Center for Firm and Organization Studies (CFOS) of Renmin University of China, which gave me a lot of help through Young's Workshop and CFOS's seminars. I would also like to thank my doctoral tutor Prof. Ruilong Yang, who has given me many constructive suggestions for my research on this project and due concern to my academic career development. I would also be indebted to academic colleagues, Prof. Naiwu Hu, Prof. Ye'an Zhou and Prof. Qijing Yang for their support and guidance.

Finally, I would like to thank my friends and colleagues who work in the political community and enterprises for their first-hand materials and my family for their patience, encouragement and constant support.

Contents

Part II The Impact of Collusion and Institutional Environment

List of Figures

List of Tables

List of Maps

PART I

An Anatomy of the China Model from the Perspective of Collusion

CHAPTER 1

Introduction

1.1 The China Model

China, which although has slowed down its economic growth rate over the past 5 years, has actually been zipping along its way to modernization since 1978 when it commenced the policy of reform and opening-up to the outside world. Its Gross Domestic Product (GDP) increased, on the average, by approximately 10% annually.[1] According to the comparable statistics (calculated in terms of the US dollar in 2000) provided by the World Bank, I draw a GDP growth chart (see Fig. 1.1), which is designed to compare the GDP growth rates among the following seven major economies: USA, UK, Japan, Germany, France, Canada, and China. It is easily noted from Fig. 1.1 that China led the other major industrialized countries by a large margin in economic growth. Devastated by the 2008 American subprime crisis and European sovereign debt crisis (SDC), each economy in this figure suffered the steepest annual decline in 2009, when all economies, excluding China, underwent minus growth. Japan's economic growth rate plummeted by 6.3%. Despite a slightly slower growth rate than previous years, China still maintained an enviable rate of 9.2% in 2009.

Prominent as it is in the garden of economic growth, China is on no account the only resplendent flower because some other burgeoning economies such as BRICs (Brazil, Russia, India, and China) also cut a conspicuous figure. Similarly, according to the comparable data from the World Bank, I also draw a chart (see Fig. 1.2) which shows the growth

H. Nie, *Collusion, Local Governments and Development in China*,
DOI 10.1007/978-981-10-5059-6_1

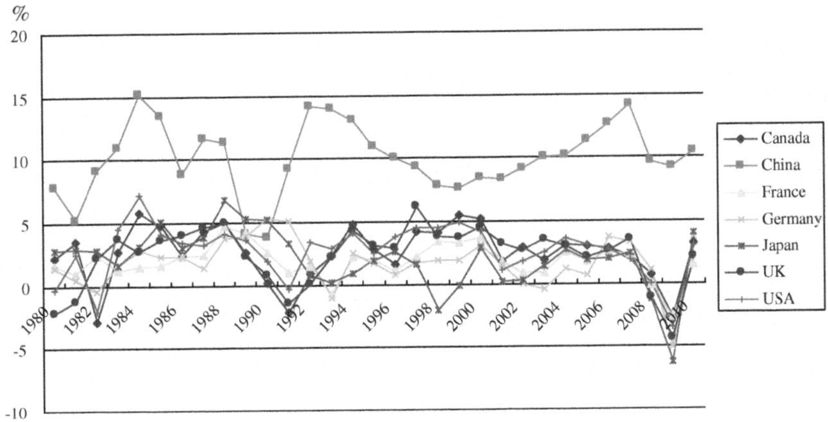

Fig. 1.1 GDP growth rates of major economies (1980–2010)

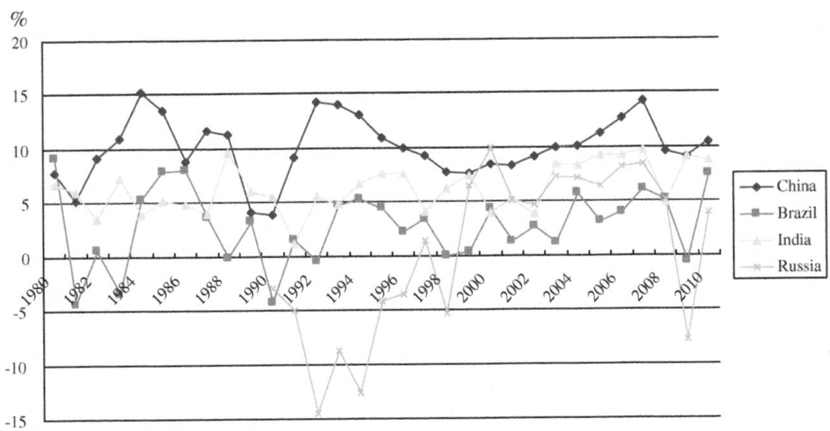

Fig. 1.2 GDP growth rates of BRICs (1980–2010)

rates of the BRIC countries in the three decades from 1980 to 2010. Even if compared with other exuberant emerging economies, China still took the lead and was overtaken only in sporadic years by Russia and

India. If compared with industrialized economies in Fig. 1.1, Brazil, India, and Russia are all riding on the rising momentum and their potential growths are promising. As a matter of fact, China's neighboring countries, such as Vietnam, Indonesia, and Thailand, by taking advantage of their excessive and cheap workforce, have replaced China as the "workshop of the world" in some sectors of manufacturing. According to the World Investment Report 2012 released by the United Nations Conference on Trade and Development (UNCTD), Southeastern Asian countries attracted \$117 billion foreign direct investment (FDI), increasing by 26% when compared with 2011. However, China attracted only \$124 billion, and the growth rate dropped to only 8%.

It is an indisputable fact that China has made a great achievement in economic construction. In recent years, some economists dubbed the experience China has acquired in economic development as the China Model. *The Economist* claimed that the China Model has taken the place of the earlier Beijing Consensus and become a new development model.[2] The Beijing Consensus has ever been regarded as the opposite to the Washington Consensus. Therefore, a thorough understanding of the China model entails an insight into the Beijing Consensus and the Washington Consensus.

The concept of Washington Consensus was put forward by economist John Williamson (1989) to refer to a set of 10 relatively specific economic reform prescriptions that, according to his interpretation, constituted the standard reform package promoted for crisis-ruined developing countries by Washington, D.C.-based institutions such as the US Treasury Department, the World Bank, and the International Monetary Fund (IMF). The consensus as originally stated by Williamson (1989) included the following 10 sets of relatively specific recommendations:

1. Fiscal policy discipline, with avoidance of large fiscal deficits relative to GDP;
2. Redirection of public spending from subsidies ("especially indiscriminate subsidies") toward broad-based provision of key pro-growth, propoor services like primary education, primary health care, and infrastructure investment;
3. Tax reform, broadening the tax base, and adopting moderate marginal tax rates;
4. Interest rates that are market determined and positive (but moderate) in real terms;

5. Competitive exchange rates;
6. Trade liberalization: liberalization of imports, with particular emphasis on elimination of quantitative restrictions (licensing, etc.); any trade protection to be provided by low and relatively uniform tariffs;
7. Liberalization of inward foreign direct investment;
8. Privatization of state enterprises;
9. Deregulation: abolition of regulations that impede market entry or restrict competition, except for those justified on safety, environmental, and consumer protection grounds, and prudential oversight of financial institutions;
10. Legal security for property rights.

The term Beijing Consensus was coined by Joshua Cooper Ramo, a former senior editor and foreign editor of *Time* magazine and later Vice Chairman at Kissinger Associates, to pose China's economic development model as an alternative to the Washington Consensus. Ding (2011) described the Beijing Consensus as consisting of the following four points:

1. ban on indiscriminate privatization of enterprises in all sectors;
2. restriction on quick privatization of financial and banking services;
3. prohibition of free international trade;
4. adherence to China's own blueprint of political and economic development and negligence of the suggestions from the US-led Western countries.

However, the past three decades since the commencement of reform and opening-up witnessed the acceleration of marketization and deregulation of economic and social restrictions.[3] To do it justice, China's development reflects more recommendations encompassed in the Washington Consensus than the points contained in the Beijing Consensus. As a matter of fact, it was the combination of the recommendations of the Washington Consensus with China's specific national situations that propelled its reform of tax-sharing system in 1994, the strategic adjustment of state enterprises after 1999 and its entry into the WTO in 2001. Thus, the approach to economic development on China's basic national conditions was dubbed as the China Model. In an article carried in *The American*, Callic sums up the China Model as "economic freedom plus political restrictions," among which economic freedom is the quintessence of the Washington Consensus. Though the term "the China Model" has gained popularity among the economic and political communities, recent economic literature (Yao 2008; Zheng 2010; Huang 2011; He and Lu 2012) shows that there has never been a consensus as to its definition, and some scholars even doubted whether there exists a "China Model." If "model" is defined, in this particular context, as a steady, mature, and reproducible development road, a China Model does not exist at all for the following two reasons:

1. China is a special country, which is quite different from most developing countries in population size, political and economic systems, topography, and environment.
2. China is still exploring a development road suitable to its own peculiar situations. Furthermore, China still has a spectrum of problems which pose an impediment to sound, stable, and sustainable development.

However, a China model, if formulated on the basis of an analysis of the characteristics of China's quick economic development over the past 30 years, will be acceptable. In its narrow sense of economic development approach, the China Model can be summarized as governmental guidance, investment motivation, demographic dividend, extensive economic growth, dependence on export (export substitution), and state holding. In its more broad sense, Ding (2011) expounded the three interwoven subsystems: Leninist political framework, social control system with Chinese characteristics, and government-regulated market economy.

This book focuses not on the controversial concept of the China Model but on the causes and impact of China's economic growth and prospects of economic reform. The controversy over the definition of "the China Model" will not prevent us from having a comprehensive research into the path China has covered in its economic and social construction and summing up gains and losses, advantages and disadvantages so as to improve social well-being. This research attempts to explore the underlying reasons for, and mechanism and effects of China's extraordinary economic growth from an objective, comprehensive, and reasonable perspective.

1.2 Cost of Economic Growth

The robust and enviable economic growth is only part of the story about China's success. An attempt to have a panorama of this super economy will inevitably make us have an understanding of the price which it has paid for speedy economic growth, i.e., the cost of the China Model.

First, for all slight improvements in coalmine safety assurance in recent years, the high frequency of coalmine accidents have commanded a great concern of the world because these mishaps have become the culprit of coal workers' safety and health. In 2004, China's coal production accounted for 40% of the world total,[4] but its coalmine fatalities accounted for 80% of the global total.[5] Coalmine accidents incurred not only much damage to properties but also great loss of many lives. According to the database which our research team made, 98% of coalmine accidents incurred loss of lives and each accident claimed 2.3 lives on average. According to official statistics, from 1949 up to now, eight of all the coalmine mishaps were rated as "extremely disastrous accidents," each killing more than 100 miners. Six of these accidents happened after 2000 and the deadliest one killed 214 miners. China's coalmine safety records were much worse than those of the USA. Table 1.1 is a comparison of coalmine fatalities between China and the USA. Between 2004 and 2010, about 4090 miners were killed in coalmine disasters in China

Table 1.1 A Comparison of Coalmine Fatalities in China and USA

Year	*China*	*USA*
1999	6 399	NA
2000	5 798	NA
2001	5 670	NA
2002	7 462[a]	NA
2003	6 177	NA
2004	6 027	28
2005	5 986	33
2006	4 746	47
2007	3 598	34
2008	3 210[a]	30
2009	2 631	18
2010	2 433	48

Sources www.chinasafety.gov.cn and http://www.msha.gov.cn
Note Due to omissions, the Fig. 7462 in 2002 is taken as half of 14,924, the total number of people killed in industrial and mining industries; the Fig. 3210 in 2008 is calculated according to Xinhuanet and the Coal Industry Statistical Bulletin in 2008

on average every year while only 33 in the USA. The coalmine death rate (calculated on the basis of one million tons of coal production) in China was dozens of times more than that in such developing countries as India and South Africa (Wang 2006).

Second, considerable amount of energy consumption has brought about a serious adverse impact on the environment. Particularly, the air quality in China's urban areas is very bad and is still worsening. According to *The New York Times*, approximately only 1% of all the cities in China pass the EU's basic air quality standards.[6] In May 2011, the US Embassy in Beijing released its air monitoring data, i.e., air quality index based on the density of fine particulate matter (PM2.5), which triggered a heated debate among online communities. In addition, water pollution is also a serious problem. In 2010, two Canadian researchers Aaron van Donkelaar and Randall Martin at Dalhousie University, Halifax, Nova Scotia, Canada created a global map of PM2.5 (see Map 1.1) by blending total-column aerosol amount measurements from two NASA satellite instruments with information about the vertical distribution of aerosols from a computer model. This map clearly shows that air pollution has been an unusually austere problem in China, especially in North China and East China.[7] In addition, approximately a quarter of Chinese people cannot get clean drinking water, and over 70% of rivers and lakes are polluted (Jia 2012).[8]

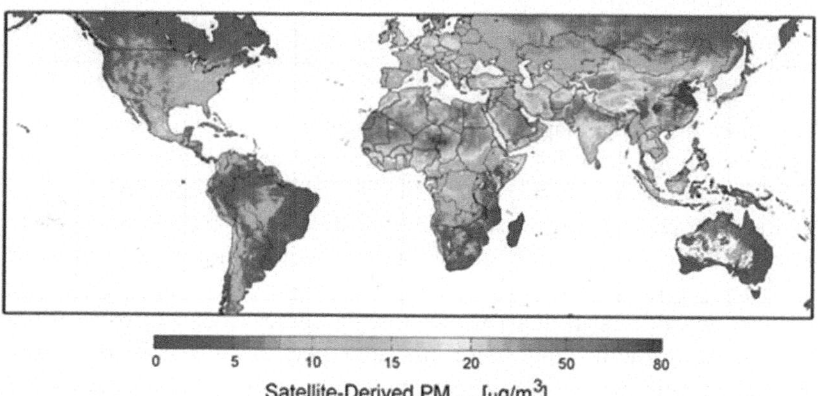

Map 1.1 Satellite-Derived PM 2.5. *Note* This map is drawn by Aaron van Donkelaar and Randall Martin, https://www.nasa.gov/topics/earth/features/health-sapping.html. (Accessed May 18, 2011)

Third, food safety is another challenge that confronts the Chinese government. In September 2008, Sanlu Group Co. Ltd. was reported to produce an adulterated milk powder contaminated with melamine, a poisonous chemical substance which can cause kidney stones and other complications. This scandal affected about 294,000 Chinese infants and killed 6.[9] Besides melamine, other chemical substances were illegally adulterated into food to the degree that the Internet was 'flooded with reports of poisonous rice, leather milk, swill-cooked dirty oil, rotgut adulterated with wood alcohol, expired steam buns dyed with forbidden coloring, chemical hot pot, fish bred with contraceptive pills, jelly mingled with industrial gelatin, salted duck eggs with Sudan dyes, lean-meat-powdered ham, etc.

In addition to coalmine accidents, environment pollution, and food safety, a wide spectrum of problems plagued this economic power. The problem of exorbitant house prices is still bothersome to many urban residents and collegiate graduates. Many jerry-built projects, including buildings, bridges, and superhighways, seriously damaged the governmental images. Counterfeit and shoddy products disrupted the market order. What is worse, mass incidents in some regions triggered public conflicts. Statistics show that there were about 8700 mass incidents in 1993, about 10,000 in 1994, 60,000 in 2003, 87,000 in 2005, and 90,000 in 2006 (Liu 2011). All the above problems are concomitant with China's economic development. Therefore, a comprehensive view of the China model entails a down-to-earth analysis of the underlying reasons.

1.3 Main Arguments

Why have accidents happened in an unending flow in the course of China's economic development? Why does the central government periodically regulate production activities of enterprises—rectifying problematic sectors, cleaning up polluting projects, and bludgeoning house prices? What is the way out for the high cost of economic development? All these questions arouse the interests of some economists. This book attempts to address itself to the following three tasks: (1) to expound "high-speed growth" and "high-frequent pernicious accidents" within the unified framework of government-firm collusion; (2) to analyze the far-reaching effect of the China model on social development; and (3) to investigate the institutional root of this model and attempt to point out the outlet for reform.

To complete the first task, we design a triple game model of "Central Government—Local Governments—Firms" and present a framework of collusion between local governments and firms which can explain high growth rate concomitant with a high frequency of pernicious accidents. Imagine, in the process of economic development are three game participants: the central government (principal, P), the local governments (supervisor, S) and firms (agent, A). Under the Chinese-style decentralization, the central government delegates its authority to local governments (at the level of provinces, municipalities, and autonomous regions) which, in turn, supervise corporate activities, develop local economy (measured by GDP), and pay a certain proportion of the tax revenues to the central government (much similar to the 1994 tax-sharing reform). This is a grand contract among the above three participants. A firm can take two modes of production: a good one with high cost but low frequency of accidents and a bad one with low cost but high frequency of accidents.[10] The central government can only observe the economic output but does not know exactly what technical means a firm utilizes, but local governments know well and even decide technical options for firms. Under the condition of information asymmetry, officials of local governments collude with firms, allowing them to choose a "bad" mode of production so that officials could be economically rewarded and politically promoted to a higher position. When colluding, local governments and firms sign a side contract whereby local governments get the bribery and whitewashed administrative performances or "political achievements" (or rather *Zhengji* in Chinese *pinyin*) and the firm rakes in much money because of their low-cost operation—such as abominable working environment, increased labor intensity, inferior raw materials, shoddy products, etc. A bad mode of production will consequently incur a wide range of accidents, which are very likely to be disclosed by the fourth party (such as mass media). If local governments and firms violate relevant regulations, the central government will inflict on them penalties, such as removing officials from office, closing firms, or imposing a fine on them. The model can be illustrated in the following figure:

Conclusion 1: *If the expected yield of the collusion between local governments and firms exceeds the expected cost, the two parties will collude to choose a bad mode of production, which inevitably breeds attendant accidents in the process of increasing economic aggregate.* To be specific, local governments and firms will risk choosing a bad mode of production on the condition of substantially tempting returns—including the increase of GDP aggregate and government revenue in its short tenure, likely promotion to a higher position because of prominent political achievements in the inter–local–governmental competition,[11] and a breathtaking sum of perks and bribes offered by the favored firms. However, this low-cost mode of production is responsible for production safety accidents, including the above-mentioned coalmine accidents and serious environmental pollution. High growth rate and numerous accidents are the unavoidable consequence of the collusion between local governments and firms.

Conclusion 2: *The central government should prevent local governments and the firms from colluding. It should enforce harsher penalties local governments and firms which violate rules, strengthen the surveillance function of the fourth party (mass media and public opinions), and encourage corporate technological innovation.* Intuitively, to diminish or even eliminate, collusion is equated with decreasing its revenue or increasing its cost. Given that the comprehensive performance assessment system remains unchanged, increasing the cost of collusion is the only means to deter collusion. Enforcing stricter punitive measures on firms for their violation of rules and dereliction of duty can reduce their expected revenue from collusion. Strengthening supervision of the public opinion and especially utilizing the current media—blogs, microblogs, and BBS etc. can effectively better the information structure of the central government, reduce the possibility of information asymmetry, and improve the supervising and ruling ability of the central government. In addition, if the central government encourages the firm to make renovations, such as reducing the cost of good mode of production and improving the market efficiency of good technology, there will be no motivation for a firm to risk for more profits.

Now that theoretically the central government can design an optimal mechanism to prevent collusion, why realistically does the central government permit collusion between the local government and the firm? Or rather why do a spectrum of problems such as production safety accidents, environment pollution and housing bubble ceaselessly happen? To answer the question, we must temporarily put the second task aside and solve the third task: investigating the institutional root of this model and attempting to point out the way out for reform. Suppose the central government has two objective functions—economic growth and social stability—and the two objectives are conflicting. In this case, the central government faces a tradeoff. The increase of social wealth as a result of speedy economic growth consolidates the legitimacy of the ruling party and the authority of the central government, but the widening gap between rich and poor as a result of high economic growth jeopardizes social stability and hence undermines the authority of the central government. Therefore, the central government will by no means go to extremes. Instead, it will weigh economic growth against social stability and readjust its attitudes and policies concerning the collusion between local governments and firms.

Conclusion 3: *Given the tradeoff between economic growth and social stability, the central government will permit local governments and firms to collude if economic growth is relatively important or the cost of collusion is comparatively less. Under the contrary condition, the central government will take preventive strategy against collusion.* In reality, the importance of economic growth is usually associated with the stage of economic development. In a less-developed region (medium and small-sized cities in hinterland China, for instance), economic growth is much more important than social stability in that economic growth is universally regarded as the guarantor of social stability. Quite literally, in hardscrabble regions, feeding more mouths outweighs protecting the environment. On the contrary, in wealthier regions (first-tier cities such as Beijing, Shanghai, and Guangzhou, for instance), local governments have to prioritize social stability because the problem of food and clothing has already been solved and residents naturally have a higher demand for living standards. An interesting and relevant question arises: Since the central government takes discriminative preventive measures against collusion in different regions with different stages of development, will it take differential preventive measures against collusion in the same region in different periods of time?

Conclusion 4: *The central government will periodically and alternately intervene in collusion because in a politically sensitive period when social stability outweighs economic growth, the central government tends to deter collusion and in a politically non-sensitive period when economic growth outweighs social stability, the central government permits collusion.* In China, politically sensitive periods are always associated with political conferences, such as the annual meeting of the National People's Congress (NPC) and annual meeting of the Committee of the Political

Consultative Conferences (CCPCC) at the national and provincial and county levels in the first quarter of the year and the quinquennial CPC National Congress. During these conferences, maintenance of social stability is the foremost important mission of the governments at all levels. Therefore, the central government will take strict collusion-proof measures. However, once a politically sensitive period is over, local governments shift their priority to economic construction. Of the objective variables of the central government, economic growth carries greater weight than social stability, and thus the government is prone to slacken its efforts to deter collusion. A politically sensitive period might also be linked with public opinions. Suppose, newspapers, television, and radio have saturation coverage of a coalmine accident, the central government will immediately clamp down collusion—a thorough check of production safety and shutdown of all coalmines. This can account for why the central government takes a periodical macro-control of an administrative intervention in coalmines, environment governance, and housing market. We may just as well be inquisitive: why does the central government make a trade-off between economic growth and social stability? This answer lies in a more profound level of political system and institutional environment.

Conclusion 5: *Periodical collusion will inevitably result from centralization where the central government evaluates local governments' performances in terms of GDP.* Theoretically, in view of the fact that a highly centralized authoritarian regime tends to give rise to information asymmetry between the central government and local governments, a typical moral hazard is likely to occur that the central government finds no way to have a timely monitor of local governments due to lack of perfect governance, manpower, and human resources. To reduce moral hazards, the central government cannot but base its assessment of local governments' performance on empirical observations, an alternative which only results in second best. Under the particular current circumstance, GDP, government revenue, and tax revenue are the less costly and relatively informative indicators for performance. As a result, authoritarian regime and GDP-centered evaluation system can be traced back to the same origin. The GDP-centered evaluation system entices local governments to collude with firms so as to attain his orientation goal only by choosing a less costly mode of production. When the cost of collusion jeopardizes social stability and the fundamental interests of the mass of people, the

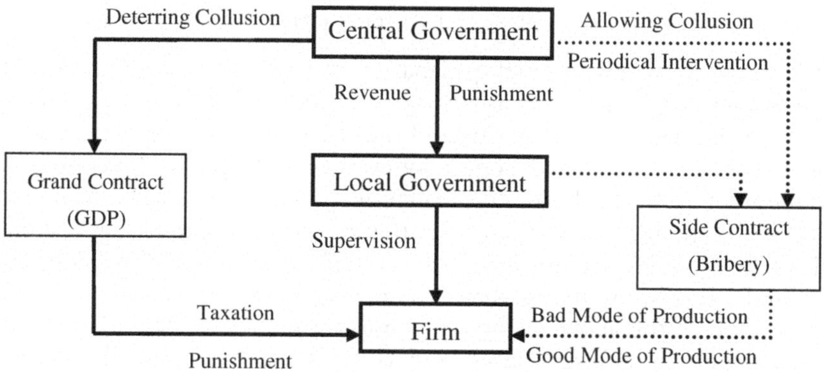

Fig. 1.3 An analytical framework of government-firm collusion

central government will deter collusion, which in turn results in cyclical collusion intervention. In the framework (see Fig. 1.3), collusion and periodical intervention are endogenous outcomes of an authoritarian regime.

As for the second task—an attempt to analyze how the collusion-motivated China Model affect economic performances and social development, I based my research on the combination of qualitative and quantitative analyses to make an investigation into the effect of collusion on many aspects such as Total Factor Productivity (TFP), technological innovation, governmental soft budget constraints, official corruption, public opinion, distribution pattern of resident income, and development of social organizations. I also made an investigation into the impact of collusion on the overall reform strategy in the next blueprint. My major findings can be generalized as follows.

Conclusion 6: *Government-firm collusion results in irrational allocation of resources, thus reducing firms' total factor productivity and demoralizing their motivation to be innovative, worsening governmental budget constraints and official corruption, weakening public opinion surveillance, widening income gap, dwarfing the development of social organizations, and increasing the cost of incremental reform for the next economic development strategy.*

1.4 Key Concepts

It is necessary to give a brief explanation of the analytical methods and clarify a few key concepts so as to make this book more comprehensible.

To begin with, it must be stressed that I try my best to make a trade-off between academic and popularity even though this is an economics monograph.[12] In this book, the analytical approaches to questions under discussion are game theory, contract theory and econometrics, and some chapters even involve institutional economics, new political economics, and a case study that is shared by management and economics. The application of the above economic methods signifies the hypothesis that governments at different levels and corporations are rational subjects whose specific objective functions can maximize their utility, efficiency, and profits. The principle for a rational man to follow is cost–earnings comparison. According to this principle, an option is made when its expected revenue surpasses its expected costs. Otherwise, this option will be discarded. Childishly obvious as it seems to be, an inquiry into what factors on earth decide the costs and earnings of a rational person requires an economist with professional knowledge to do it justice. Nevertheless, it must be clearly stated that an advanced and abstruse mathematics is totally unnecessary for a reader to understand this book. Some mathematical optimizations and general knowledge about game theory and statistical fundamentals suffice.[13]

One of the major contributions of this book is the provision of a framework of collusion between local governments and firms. This framework is effective to explain high growth rate and high frequency of pernicious accidents. It retrospects the China model from a new perspective and explains many typical examples concerning corporate and governmental behaviors, economic performances, and social development. In this book, there are three assumptions:

First, it is assumed that there exists, to some degree, a conflict between economic growth and social stability. Low-cost modes of production opted simply for boosting economic growth breeds the risks of safety misadventures and environmental disasters and, therefore, have an adverse impact on social stability. This means that there will be no difficulty in choosing a suitable mode of economic development so long as economic performances are "good and fast".

Second, it is assumed that the central government and the local government have different objective functions and therefore they diverge in the option for interests. The objective functions of the central government include economic growth and social stability, each of which accounts for corresponding weight in different periods of time. However, the local government only takes short-term (or rather near-sighted) economic returns and political rewards into consideration but neglects long-term possible aftermaths when making its objective functions because, under the Chinese authoritarian regime, top officials of local governments are usually not locals (from other provinces or municipalities or counties or townships, depending on different levels) and their tenures are usually not long.[14] In addition, local residents do not have an effective surveillance on local governments.

Third, it is assumed that there is information asymmetry between the central government and local governments while local governments share information with local firms. Supposing the central government is well informed of the local government's actions, there will be no agency and therefore collusion, but this assumption is not only of little interest in economics sense but also far beyond realistic situations.

For a reader to have a good understanding of the framework of collusion between the local government and the firm and major conclusions of this book, some keywords should be clarified as follows:

1. **Central government.** The central government in the framework, in its broad sense, refers to the Central Committee of the CPC (Communist Party of China), National People's Congress (NPC), and the State Council, all of which represent the top state organs prescribing laws and regulations. The framework is still applicable to an analysis of any collusion below the provincial level (such as prefecture, city, county, township levels and even to the lowest village) because in this case the central government in the framework actually represents any government just above the provincial or county or township level according to different situations. In principle, the framework holds true when two governments at different levels diverge when they pursue conflicting interests.
2. **Firm.** A firm in this book refers to any production unit, including any individual embarking on business or sole proprietorship. We make a sensible assumption that a firm's choice of production mode and technology is influenced and even to somewhat degree dictated by the local government which can always implement "Investment Attraction Law" to navigate the firm to make a decision and use authorized power to approve or veto the firm's choice.

3. **Information asymmetry.** In game theory, one participant may have more and better information (cost, price, policy, etc.) than other parties and this creates an imbalance of power in transactions. In the case of information asymmetry, the party who has more and better information is likely to tell a lie and loaf on the job.
4. **Accident.** An accident in this book refers to any event jeopardizing social stability. It includes coalmine accidents, environmental pollution, food safety, counterfeit and shoddy products, smuggling, economic bubbles, and unexpected mass incidents.
5. **Collusion.** This is the core concept of the monograph. For the sake of their own interests, local governments violate laws and regulations formulated by the central government and collude with firms to make a choice contrary to relevant regulations. It is necessary to draw a distinction between collusion and some relevant concepts to guarantee a comprehensive and precise elaboration of the core concept. First, collusion is different from political connection. The latter often refers to the relationship between an individual official and a firm and their relationship is an exchange of interests (Fisman 2001; Faccio 2006). However, collusion is not merely concerned about the relationship between an individual official and a firm but also about the relationship between the local government and the firm. An official may collude with a firm simply for fishing for personal gains, but the governmental regulations represent the will of the collective. For instance, to boost economic growth, a local government (collective) introduces some polluting enterprises from which no single official (individual) may get any profit. Second, collusion is not an equivalent to corruption. Corruption refers to as an official's abuse of public power (illicit conduct) for his own personal interests. An official's choice of a bad mode of production for the purpose of gaining both political and financial interests is a substantial manifestation of collusion, whereas only the forms of corruption closely associated with a local government's cash-out of political achievements can be counted as a substantial manifestation of collusion. However, as mentioned above, collusion is not always corruption or illicit conduct. Third, although collusion is roughly equal to "official-coalminer collusion" often quoted by mass media, this book attempts to explain the reason for "concurrence of high growth rate and high frequency of pernicious accidents" in terms of "local

government-firm collusion," a concept specifically referred to as a local government's deliberate option for bad modes of production simply for political gains.

6. **Whipping top equilibrium.** This term is coined by the author of this book. China's economic operation is just like a whipping top which, but for its high-speed gyration, cannot maintain its balance. Once its gyration slows down, it will gradually fall down. The reason why I compare China's economic operation to whipping top equilibrium is that high-speed economic growth brings about two contradictory consequences: alleviating and even eradicating poverty and satisfying people's increasing demand for wealth on the one hand, but concomitantly incurring a broad range of pernicious accidents on the other. However, if economic growth rate sustains very low, many serious social problems—unemployment and poverty and possible social upheavals in the wake of them, for instance—will inevitably arise. As far as its social system and productivity are concerned, it is difficult for China to enter into an ideal period of "high growth rate and low frequency of pernicious accidents." There is unlikelihood for China to maintain a high economic growth rate all the time and this is the potential crisis of the Chinese economy. As a result, China should improve its institutional quality through reform so as to get rid of whipping top equilibrium.

1.5 The Structure of This Book

This book is divided into two parts. The first part, including five chapters, explains "high economic growth rate and high frequency of pernicious accidents" through the frame of collusion. The second part, also including five chapters, analyzes the impact of collusion on economic and social development and attempts to explore the institutional root and the feasible way out.

The present chapter (Chap. 1) is an introduction which teases apart the relations of all the chapters in this book so as to help readers have a better understanding of it.

Chapter 2 provides a theoretical model for this book—the framework of collusion. This chapter is somewhat abstruse because it contains an analytical framework, verifiable propositions, and the implications of governmental policies. Therefore, a good understanding of this chapter is the key to understanding the whole book. The students and professionals interested in collusion, if trying to develop the theoretical model, might have an insightful notion on it.

Chapters 3–5 apply the framework of collusion to analyzing coal industry, housing sector, and other sectors, respectively, all of which are carefully chosen. Chapter 4 is illuminating in that it provides not only some criteria to testify to the existence of collusion but also an empirical approach to collusion. Chapter 5 makes a tentative analysis of environmental pollution, food safety, tax evasion and tax fraud, jerry-built projects, and smuggling through a series of case studies within the framework of collusion.

Chapter 6 discusses the impact of collusion on corporate behavior and corporate performance, including technological innovation, Total Factor Productivity (TFP), and Foreign Direct Investment (FDI). Chapter 7 analyzes, with a case study of Saiwei LDK Solar Co. Ltd—a manufacturer of photo Voltaic (PV) products, solar wafers and modules in Jiangxi Province in Central China, the impact of collusion on official corruption and governmental soft budget constraints. Chapter 8 discusses the impact of collusion on opinion surveillance, income distribution, the development of social organization, and the current overall strategy of reform.

Chapter 9 explores the institutional environment of collusion. Not only does it analyze the inevitability of collusion or cyclical collusion from the perspective of centralization, political target, legal environment, and local governmental competition but also provides a tentative computational test.

Chapter 10 retrospects the sustainability of the China Model, believing that collusion is a unique "whipping top equilibrium." Only when China gets rid of demographic dividend on which the China Model has relied and gains institutional dividend by improving institutional quality, can China be a real world power. Then it discusses the feasible way out to follow when we make a shift from collusion to cooperation. The last section of this chapter summarizes the shortcomings of this book and points out the topics deserving further research.

Notes

1. This figure is calculated according to the data provided by the World Bank, http://data.worldbank.org/ indicator (Accessed June 28, 2009).
2. "The Beijing Consensus Is to Keep Quiet", *The Economist*, May 6, 2010. http://www.economist.com/node/ 16059990 (Accessed January 18, 2011).

3. Yao Yang (姚洋), professor of National Development and Research Institute in Beijing University, pointed out in a lecture that China's progress in economic growth is a case in point of the Washington Consensus. http://old.ccer.edu.cn/cn/ReadNews.asp?NewsID=9415 (Accessed May 18, 2011).
4. BP Statistical Review of World Energy, http://www.bp.com/statistical-review (Accessed December 15, 2005).
5. "Coal Mine Fatalities in China Account for 80 Percent of World Total", BBC, November 13, 2004.
6. Kahn, Joseph and Jim Yardley. "Choking on Growth: Part I-IX", *The New York Times*, August 26, 2007.
7. See "NASA releases global air-quality map and East Asia is heavily polluted". http://news.sina.com.cn/green/news/roll/2010-09-27/140321179715.shtml. See also: http://english.cntv.cn/special/pm25/Homepage/ (Accessed May 18, 2011).
8. Jia (2012) provided more background information about the environmental pollution in China.
9. Beach, Sophie. "Why China Struggles with Food Safety?", *The Wall Street Journal*, April 25, 2011.
10. Take electric power plants that burn fossil plants releasing hydrocarbons, nitrogen oxides, sulfur dioxides and fumes. If a power plant hopes to be up to the standard set by 'Regulations of Energy Saving and Emission Reduction' during the Eleventh and Twelfth Five-Year Plans stipulated by the State Council, it must reduce the emission of nitrogen and sulfur dioxides. However, technological improvements on the power plant inevitably increase cost of electricity generation. For instance, the costs of desulfurization and denitrification increase by 2.0 cents and 1.2 cents per kilowatt hour respectively. On the other hand, the subsidies for desulfurization and denitrification are only 1.2 cents and 0.8 cent per kilowatt hour only (see 'Can Subsidy for Electricity Price Cover Costs of Emission Reduction?' in *China Electric Power Newspaper*, December 7, 2012). This shows that environment-friendly power plants (perhaps other industrial sectors) are much less lucrative than those that fail to meet the environmental standards.
11. Competition exists between local governments at the same level (provinces, counties and townships). For example, the total of GDP aggregate reported by each province or municipality to the central government outnumbered the figure released by the National Bureau of Statistics of China. It is arithmetically ludicrous that the GDP growth rate of each province was higher than the national average. The reason is that the provincial leaders are likely to be promoted to a higher position because of their glossing luster economic achievements. This is also true to local governments at the levels of counties and townships.

12. This embodies the quintessence of economics: we are always making a trade-off.
13. It is not difficult to understand some fundamentals of economics, but only advanced economic knowledge can help us to give a reasonable explanation of economic phenomenon and to estimate whether an option is optimal or not.
14. Li and Zhou (2005) based their research on the 1979–2005 sample and found that the average tenure for top provincial leaders (party secretary and provincial governor) is only 3.03 years.

References

Ding, Xueliang. *On the China Model*. Beijing: China Social Sciences Academic Press, 2011. 丁学良. 辩论中国模式. 北京: 社会科学文献出版社. 2011.

Faccio, Mara. "Politically Connected Firms". *American Economic Review*. 2006, 96 (1): 369–386.

Fisman, Raymond. "Estimating the Value of Political Connections". *American Economic Review*. 2001, 91: 1095–1102.

He, Di and Lu Liling (eds.) *A Reflection on the China Model*. Beijing: China Social Sciences Academic Press, 2012.

Huang, Yasheng. *How Unique is "The China Model"*? Beijing: Zhongxin Press, 2011. 黄亚生. 中国模式到底有多独特? 北京: 中信出版社, 2011.

Jia, Ruixue. "Pollution for Promotion". Stockholm University, IIES working paper, 2012.

Li, Hongbin and Li-An Zhou. "Political Turnover and Economic Performance: The Incentive Role of Personnel Control in China." *Journal of Public Economics*. 2005, 89: 1743–1762.

Liu, Neng. "Mass Incidents in Contemporary China: Changes of Images and Reconstruction of Classification Framework", *Journal of Jiangsu Administrative Institute*. 2011, (2): 53–59. 刘能. 当代中国的群体性事件: 形象地位变迁和分类框架再构. 江苏行政学院学报, 2011, (2): 53–59.

Wang, Shaoguang. "Regulating Death at Coalmines: Changing Mode of Governance in China". *Journal of Contemporary China*. 2006, 15 (46): 1–30.

Williamson, John. "What Washington Means by Policy Reform". In John Williamson, eds. *Latin American Readjustment: How Much Has Happened*, Washington: Institute for International Economics, 1989.

Yao, Yang. "Is There Really a China Model?" *Reprinted Journal Materials of Renmin University, "Socialism Economic Theory and Practice"*. 2008 (10). 姚洋. 是否存在一个中国模式? 人大复印报刊资料 "社会主义经济理论与实践", 2008 (10).

Zheng, Yongnian. The China Model: Experiences and Bewilderments. Hangzhou: Hangzhou People's Press, 2010. 郑永年. 中国模式: 经验与困惑. 杭州: 浙江人民出版社, 2010.

CHAPTER 2

Government–Firm Collusion: A New Analytical Framework

This chapter is to explain the China Model characterized with "high economic growth rate concomitant with high frequency of vicious accidents" by constructing a formal model of the local government–firm collusion. Section 2.1 introduces three famous or infamous events about PX (p-Xylene) projects in China and raises two questions relevant to these events. Section 2.2 makes a literature review of theoretical interpretations of the China Model and elicits the argument of the whole chapter. Section 2.3 defines the basic elements of the model of the local government–firm collusion. Section 2.4 discusses optimal equilibrium collusion contract and optimal prevention collusion contract and testifies to major propositions of this chapter. Section 2.5 provides an interpretation of different results of the three PX projects mentioned above. Section 2.6 discusses other factors that might have an impact on collusion. Section 2.7 is a summary of the chapter.

2.1 Three PX Projects

In recent years, mass incidents came thick and fast in Dalian, Ningbo, and Xiamen where residents congregated to protest against the construction of PX projects. These incidents commanded a great concern both at home and abroad. PX, the shortened form for p-Xylene, is a colorless and aromatic yet flammable and toxic liquid mainly used for the manufacture of plastic and polyester. Its production output is a far cry from being sufficient to meet increasing demand. For instance,

H. Nie, *Collusion, Local Governments and Development in China*,
DOI 10.1007/978-981-10-5059-6_2

it was estimated by sina.com (a Website for the Chinese communities worldwide, covering a full array of news, entertainment, financial information, science and technology, and lifestyle tips) that China produced and supplied 4.8 million tons of p-Xylene in 2009, but its domestic demand was estimated to be 8.17 million tons. In 2010, China became the largest producer and consumer of this poisonous chemical, with its production capacity accounting for 20% and consumption for 30%.[1] As far as economic efficiency is concerned, the output value of a PX project (estimated to be as high as tens of billions of Chinese Yuan [CNY]) is absolutely alluring despite its huge cost of construction (estimated to be more than CNY10 billion[2]). At present, in Mainland China, 15 PX projects have been completed or are still under construction. Nearly all of them are concentrated in the eastern coastal provinces.

2.1.1 *Dalian XP Project: Relocation and Expansion*[3]

The Fujia-Dahua PX project in Dalian, a well-known deepwater seaport city in Liaoning Province, Northeast China, was approved by National Development and Reform Commission (NDRC) in December 2005, and it was put into production in June 2009. Fujia Dahua Co. Ltd. was co-founded by Dahua Group (underwriting CNY49 million), a Dalian-based holding company under the control of the Dalian Municipal Government and Fujia Investment Co. Ltd. (underwriting CNY51 million), a private enterprise.

Devastated by a disastrous typhoon *Muifa* on August 8, 2011, the breakwater protecting the Fujia-Dahua plant collapsed and the surging seawater rose near p-Xylene storage tanks. Although it was officially claimed that later the risk of p-Xylene leakage had been removed, the company's prevention of journalists from making a field investigation was widely spread within the national mass media. Furthermore, the media disclosed a bombshell that the plant's practical operation preceded its completion acceptance. On August 14, tens of thousands of Dalian dwellers congregated on the People's Square and protested against the PX project. The Dalian Municipal Government announced that night that Fujia-Dahua should immediately discontinue its production and decided to relocate the project. But no one knew when and where the project would be relocated.

In August 2012, a year after the Dalian Municipal Government's announcement of the project relocation, *Dalian Daily* reported that the relocation plan had passed the review conducted by the authoritative

China International Engineering Consultant Corporation Limited. According to the plan, Fujia-Dahua would be relocated from Dagushan petrochemical enterprise complex to Changxing Island Industrial Zone. On October 29, 2012, people.cn, a news platform created by China's most influential newspaper *People's Daily*, released an article titled "Fujia-Dahua PX project still in operation, capacity doubled to 1.4 million tons". In late December 2012, *Shanghai Business Daily* quoted from a Dalian official that the PX project was up to the standard after a thorough rectification and "production has been resumed recently".

2.1.2 *Xiamen's PX Project: Relocation*

In February, the State Council approved the Haicang PX project financed by a Taiwan-funded enterprise Dragon Aromatics in Xiamen (Amoy), a famous seaport city directly under the jurisdiction of Fujian Province, Southeast China. The factory, with an estimated annual production of 700,000 tons of p-Xylene would be constructed in Haicang District. According to the original plan, the construction project commenced in November 2006 and it was put into production in 2008.

In March 2003, Zhao Yufen, an Academician of the Chinese Academy of Sciences, and other 105 members of the National Committee of the Chinese People's Political Consultative Conference initiated a draft resolution "Proposal about the relocation of Haicang PX project", but later this proposal failed to get approval.

On May 30, 2007, the Xiamen Municipal Government resolved to delay the PX project. The Fujian Provincial Government demanded the Xiamen Municipal Government to make a more comprehensive environmental appraisal about the project.

The next day, thousands of Xiamen residents had a demonstration before the Municipal Government Hall. In December 2007, on the basis of environmental appraisal, Internet vote and two rap sessions attended by city residents, both the Fujian Provincial and the Xiamen Municipal Governments decided to abide by public opinion and relocate the PX project to Gulei Peninsula, Zhangzhou.

On January 9, 2009, the National Environmental Protection Department approved in principle the PX project of Dragon Aromatics (Zhangzhou). The project's total investment amounted to CNY13.78 billion, of which CNY0.83 billion was earmarked for environmental protection. The annual discharge amount of sulfur dioxide was 3644 tons and annual chemical oxygen demand (COD) was 56.16 tons.

2.1.3 *Ningbo's PX Project: Abortion*

In the Twelfth Five-Year Plan (2011–2015), the Ningbo (a seaport city under the jurisdiction of Zhejinag Province, East China) Municipal Government proposed to introduce a series of petrochemical refinery and integration projects (including a PX project). The prospective site would be situated in the Economic and Technological Development Zone, Zhenhai District, Ningbo. It was estimated that the total investment would be CNY55.873 billion. If it were completed, Zhenhai Refinery & Chemicals Engineering Company would be the largest complex oil refinery enterprise in Asia.

In early October 2012, some villagers near the would-be construction site, through the Public Voice System, complained with the Zhenhai District Government against possible environmental disaster and requested to relocate their homes. On October 22, nearly 200 villagers also filed their complaints with the Ningbo Municipal Government. Since then, a legion of local villagers and city dwellers consecutively expressed their resentment, appealing to the authorities to halt the project.

On October 27, the Ningbo Municipal Party Committee and the Ningbo Municipal Government held a symposium, making a comprehensive survey of the project. The next day, under the pressure of continuous massive demonstrations, a spokesman of the Ningbo Municipal Government expressed explicitly: "Ningbo will resolutely give up on the PX project; the petrochemical refinery and integration project will be immediately discontinued until a scientific research testifies to its feasibility."

Two questions are elicited from the above three cases about the PX project: (1) Why did some local governments defy resistance and pressure from local residents and doggedly push forward p-Xylene projects? (2) Why were some projects aborted while others were put into operation smoothly or relocated?

2.2 The China Model Driven by Government–Firm Collusion

It is necessary to make a general review of recent literature concerning the China economic growth model before we make a research on the China Model from the perspective of government–firm collusion. The China model is two-fold: high economic growth rate concomitant with high frequency of vicious accidents. To do it justice, the publications as to the China Model explain well either side (positive or negative) of the model, but none of them analyzes within the same framework both sides of it.

Large quantities of literature ascribe high economic growth to the local governments and especially underscore the role they play in competing for GDP aggregate (Yang 1998; Zhang and Su 1998). Qian and Xu (1993) and Maskin et al. (2000) argue that the Chinese authoritarian regime is a multidivisional government structure where each division, which is very much like a business department within a gigantic business corporation, has a similar industrial structure. That is why local governments can carry out yardstick competition. But why do local officials have the incentive to participate in GDP competition? Some researchers believe that local officials could gain from faster economic growth more financial revenues since the implementation of local financial contract system in the early 1980s and tax-sharing system in 1994 (Oi 1992; Lin and Liu 2000). In terms of contract theory, the gain-sharing contract crafted by the central government as the principal can alleviate moral hazards of the local government as the agent. Unlike Oi's Local State Corporatism, this type of political centralization and economic decentralization is labeled as market-preserving federalism (Weingast 1995; Qian and Weingast 1996; Jin et al. 2005).[4] Other researchers observe that government officials with better economic performances are more likely to be politically promoted to a higher position. Bo (1996) makes an empirical study on the political promotion of all provincial governors and provincial vice-governors since the founding of the PRC in 1949 and draws a conclusion that their promotion is linked up with their economic achievements, especially with the tax revenue paid to the central government. Li and Zhou (2005) and Chen Ye et al. (2005) draw a similar conclusion.[5] However, only a few articles accentuate the costs incurred by competitions between local governments. These costs include: local protectionism and repetitious construction (Shen and Dai 1990; Zhou 2004; Bai et al. 2004); local governments' collusion with local firms to evade taxes or escape administration, thus undermining the central government's monitoring capacity (Cai and Treisman 2004); local governments' conflicting multitasks—such as a vanity project versus a livelihood project (Nie and Li 2006) and the widening developmental gap between regions (Zhang et al. 2007).

Obviously, cost and revenue in the process of China's economic growth are the two sides of the coin of the China model. However, as far as I can tell, there is no article or monograph which can explain high economic growth rate concomitant with high frequency of vicious accidents. Much literature focuses on the one side and just mentions the other side through extended discussion. From the perspective of economic research, a self-consistent theory must get its own cost and revenue from the same set of assumptions and then decides the optimal option. In its realistic sense, a story of the China Model must explain why its high economic growth rate is concomitant with high frequency of vicious accidents.[6] By introducing a sound mathematical model, Nie and Li (2006, 2013) explain, within a government–firm collusion framework, why this concomitance occurs in China. This chapter is mainly devoted to the political–economic model of government–firm collusion.

This model has three game players: the central government (principal), local governments (supervisor), and firms (agent). Since a local government is authorized by the central government to supervise a firm's production activities, the principal knows little about the firm's cost production but the supervisor knows a thing or two. Under the situation of information asymmetry, the local government colludes with the firm to choose a bad and inexpensive mode of production over a good and expensive one for the purpose of accelerating economic growth and thus earning more profits by reducing cost input. However, a bad and inexpensive mode of production is highly likely to incur vicious accidents. Therefore, the local government–firm collusion is the root of the concomitance of high economic growth with high frequency of vicious accidents. This book presents an optimal equilibrium-collusion contract, where the central government will turn a blind eye to the collusion when the collusion-proof costs surpass the prospective advantages. This book also offers an optimal collusion-proof contract where the central government will devise an optimal collusion-proof mechanism if the costs arising from collusion are beyond the bounds of allowed proceeds. Under the optimal mechanism, the emoluments paid to the local government and the firm is endogenous, so are the punitive measures against them. Finally, we take the PX project as an example to show that the main conclusions of this chapter have been supported by empirical evidence.

The government–firm collusion model makes the following contributions to existing literature.

First, it offers a new perspective from which to analyze the China Model and thus have a panorama (analyzing both cost and proceeds) of China's economic growth. A large amount of recent literature—such as Young (2000), Chow and Li (2001), Cheung (2008), Brandt and Rawski (2008)—attempts to offer a reasonable explanation of China's high economic growth. Unlike the literature mentioned above, this book lays equal emphasis on the successful side of economic growth and on China's economic decentralization and cost of local competition.

Second, it offers a microeconomic basis on which to understand economic growth in terms of information asymmetry. Some literature reveals that local competition will generate negative effects, such as reduction in public goods and relaxation of control environment (Gordon 1983; Oates and Schwab 1988; Bandhan and Mookerjee 2000) and other literature analyzes the weakening of the central government's authority and governing competency due to local competition (Cai and Treisman 2004). However, the driving force of the above literature is the negative externalities of local competition; but the driving force of the model in our book is information asymmetry, and moreover, it endogenously explains why the central government allows collusion.

Third, it enriches the theoretical literature concerning collusion and authorization. Since Tirole (1986), there is a large set of mechanism design literature on collusion, such as Kofman and Lawarree (1993), Laffont and Martimort (1995), and Faure-Grimaud et al. (2003). The literature above discusses how to prevent collusion while the model in this book discusses both collusion-proof measures and equilibrium collusion.[7] The classical collusion model is an adverse selection model, but this model is a moral hazard model which implies that the agent's action is endogenous. In addition, this chapter can be considered as an application of equilibrium collusion and collusion-proof model to the environment of China's political economy. This chapter shows, through a three-tier (principal-supervisor-agent) analytical framework, that the Chinese-style authoritarian regime characterized by the combination of political centralization and economic decentralization can also be analyzed within the framework of a classical contract theory.

The research reveals that the GDP- or financial revenue-oriented Performance Assessment System for government officials should be reformed so as to remove the negative externalities resulting from local competition.

2.3 Government–Firm Collusion Model[8]

2.3.1 Game Players

Consider a three-tier vertical structure composed of a principal, a supervisor, and an agent and they are respectively represented by the central government, a local government, and a firm. Suppose among them there is an official grand contract by which the central government authorizes the local government to supervise the firm's production activities. The local government plays a central part in this structure because the central government is unlikely to directly oversee the production activities of all firms due to the limitations of time and information. As a production unit, the firm pays taxes to the central government. In reality, the central government refers to all central organs which are authorized by the constitution or other relevant laws to exercise macro-regulation over the production activities of all firms. In a broader sense, the local government refers to any of the CPC's committees, the executive branches, and legislative People's Congresses at the levels of provinces, counties, and townships and, in addition, other agencies such as administrations of work safety, of quality supervision, inspection, and quarantine, and bureaus of environmental protection.[9] The firm is engaged in production activities within the range of permissions prescribed by the central government and gets economic surpluses. In reality, the principal–supervisor–agent relationship exists not only in production activities but also in the provision of public goods.

Suppose the central government's target output is x (measured in terms of GDP[10]), which, in this context, is standardized as 1. The central government imposes tax t out of the total 1 on the firm, and reserves part of it as rewards $w = \alpha t$ to the local government[11] and reserves for itself $(1 - \alpha)t$, $\alpha, t \in (0, 1)$. Here, tax t denotes all sorts of taxes and tolls and αt denotes the financial revenues the local government gains. There are two modes of production c for the firm to choose: the good yet expensive one whose cost is expressed as $\bar{c}$ and under which no accidents will occur; the bad yet inexpensive one whose cost is expressed as $\underline{c}$ and under which accidents will occur with a certain degree of probability p. The verifiable *ex post* accidents, often referred to as coalmine disasters, environmental pollution, and unexpected mass incidents, incur extra social costs C, which usually include direct costs—such as loss of lives and damage to and loss of properties—and indirect costs—such as political pressures driven by public opinions (public resentment against and non-cooperation with both the central and local governments). Extra social costs undermine social stability and are eventually undertaken by the central government. Given $0 < \underline{c} < \bar{c} < 1$, $\Delta c = \bar{c} - \underline{c}$, Δc represents quasi-rent, i.e., the cost saved by the firm if it chooses a bad mode of production. A good and expensive mode the production for coalminers, for example, increases the production cost because ventilation and drainage systems and many other facilities should be installed to guarantee the safety of coalminers. According to the law stipulated by the central government, accidents are classified into different severity levels. For example, an accident with the loss of 10–30 lives is ruled as a "serious accident" and an accident with the loss of more than 30 lives as a "severe accidents with heavy fatalities." Accordingly, different punitive measures are set forth for different severity levels. As for accidents in environment pollution, energy consumption, and economic bubbles, there are different definitions of severity levels, but they do not have any slight effect on our analysis.

2.3.2 *Information Structure*

The central government does not know what production technology or manufacturing technique the firm is to opt for, but the local government does because the firm is required to get a license from the local government before it chooses a mode of production.[12] This signifies information asymmetry between the central and local governments. This firm may choose an *ex ante* good mode of production with the cost $\bar{c}$ to avoid accidents, or it may collude with the local government and choose a bad one with the cost $\underline{c}$) under which accidents may occur with a certain degree of probability p. The accidents may be reported to the central government by the fourth party (mass media) with the probability of ρ. We summarize the four possibilities as follows:

1. When the firm chooses $\bar{c}$, no or few accidents may occur.
2. When the firm chooses $\underline{c}$, accidents may not occur with the probability of $1 - p$.
3. When the firm chooses $\underline{c}$, accidents may occur with the probability of p, and the fourth party discovers the accidents with the probability of ρ.

4. When the firm chooses $\underline{c}$, accidents may occur with the probability of p, and the fourth party may not discover the accidents with the probability of $1 - \rho$.

Probabilities p and $\rho \in (0, 1)$ are common knowledge. In the possibilities of (2), (3), and (4), the local government may collude with the firm. Since accidents are verifiable hard information, the central government can conclude that the local government colludes with the firm.

2.3.3 Utility Function

The central government is risk neutral, and it simultaneously takes into consideration economic growth and social stability.[13] As a principal, the central government is the sole subject whose responsibility is to undertake all costs of economic growth and design an official grand contract stipulating for rewards and punitive measures that the local government and the firm deserve. To be more specific, the central government maximizes the total revenue $(1 - \alpha)t$ and subtracts extra social cost C that might be incurred due to the occurrence of vicious accidents.

Both the local government and the firm are risk neutral. They undertake limited liabilities[14], and their reservation utility is standardized 0. As the supervisor, the local government oversees the firm's production activities. However, the local government does not take into social cost into account because major officials are not locals and their tenure is usually short.[15] To make it simple, let us assume that the local government negotiates an agreement without giving in. This means that once the local government decides the optimal strategy, it can give the firm a "take-it-or-leave-it" offer.[16] In this case, the local government decides the firm's option for production technology and then the firm has to accept it. The local government may collude with the firm and sign a private side contract: the local government permits the firm to opt for a "bad" mode of production at the cost of $\underline{c}$ on the condition that the latter pays the former a sum of transfer payment $b = k\Delta c$ no matter whether accidents occur and any undue irregularities are disclosed by the fourth party in the future. $k \in (0, 1)$ represents the proportion of quasi-rent shared between the local government and the firm. The transfer payment may be in various forms: surcharges, recruitment of officials' relatives or more local people, the equities of the firm, and even direct bribery. Suppose this private side contract is, for both the local government and the firm, is Pareto-improving and self-enforcing (Tirole 1992). Now, we ignore temporarily the transaction cost between them.[17] The local government gets reward w, and meanwhile the firm gets $1 - t - k\Delta c - \underline{c}$ under the following two conditions: (1) not any accident occurs; (2) not any accident is disclosed even if it occurs. If the central government is informed, through the mass media or other news channels, of the local government–firm collusion, it will inflict penalties F_S and F_A, respectively, on the local government and the firm. In reality, the most common forms of penalty on local officials are dismissal or demotion. Obviously, the local government burdens only limited liabilities.

2.3.4 Time Line

This is a complete information dynamic game model. The time line is as follows:

1. The central government decides to "allow" or "deter" the local government–firm collusion, and then gives the local government and the firm a "take-it-or-leave-it" grand contract. The game continues if and only if both the local government and the firm accept the grand contract. Otherwise, the game terminates.
2. The local government decides to take either a conservative or aggressive strategy, but both of them underscore safety measures and economic growth. If an aggressive strategy is chosen, the local government and the firm collude and sign a contract.
3. The firm decides to choose good or bad mode of production. Once it chooses a bad one, it colludes with the local government and the contract is executed.
4. When a bad mode of production is chosen, accident will occur with a certain degree of probability p and may be disclosed with the probability ρ.
5. The grand contract is executed.

We can describe this time sequence as a game tree (see Fig. 2.1). U_i^j represents the expected revenue of the game participant i under the condition of j. The subscripts P, S, and A represent, respectively, principal (the central government), supervisor (the local government), and agent (the firm); the superscripts c and cp indicate collusion and collusion-proof; the underline indicates the expected revenue corresponding to the conservative strategy.

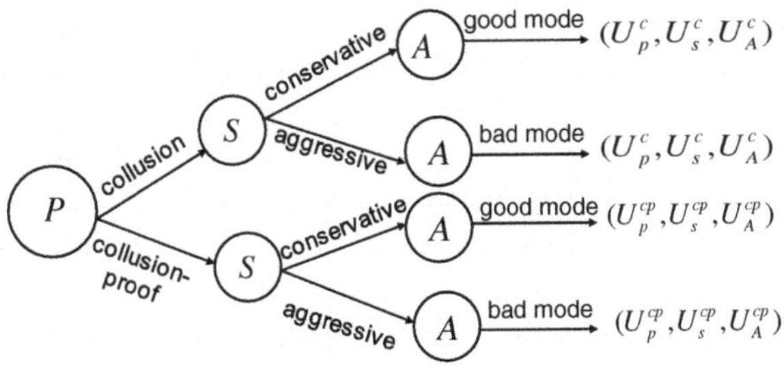

Fig. 2.1 Game tree

2.4 The Optimal Equilibrium Collusion and Collusion-Proof Contract

2.4.1 Equilibrium Collusion

The backward induction will be employed to tackle with this game. It is necessary for us to take into consideration the firm's choice before analyzing the central government's optimal strategy. As a matter of fact, the local government's choice is the foremost important key factor in that it is the supervisor of the firm. Given that the central government allows the local government to collude with the firm,[18] what is the optimal choice for the local government? Suppose the local government chooses a conservative strategy, it will demand the firm to choose a good mode of production. In this case, the expected revenues of the local government and the firm are, respectively, as follows:

$$\underline{U}_S{}^c = \alpha^c t^c \tag{2.1}$$

$$\underline{U}_A{}^c = 1 - t^c - \bar{c} \tag{2.2}$$

Contrarily, if the central government chooses an aggressive strategy, the firm will choose a bad mode of production. In this case, the expected revenues of the local government and the firm are, respectively, as follows:

$$U_S^c = \alpha^c t^c + k\Delta c \tag{2.3}$$

$$U_A^c = 1 - t^c - k\Delta c - \underline{c} \tag{2.4}$$

It is obvious that the local government's dominating strategy is aggressive if the central government allows collusion and the firm's expected revenue is not negative.[19] The reason is intuitive: collusion can bring the local government additional transfer payment $k\Delta c$. Similarly, in this case, the firm's dominating strategy is to choose a bad mode of production.[20] Because of $k \in (0, 1)$, the firm can get extra profit from the collusion. In a word, both the local government and the firm can get more revenues in the collusion than could they do without collusion. Thus, we come to the following proposition:

Proposition 2.1 *Given that the central government allows collusion, the local government and the firm will collude and adopt a bad mode of production so as to boost economy more rapidly (compared with a better mode of production) within a short period of time. In this case, the officials will have more opportunities to be promoted but the frequency of vicious accidents is also likely to be higher.*

For Proposition 2.1 to be more comprehensible, let us suppose that a province has two cities A and B. To produce 1 unit output, City A adopts a bad mode of production at the lower cost of $\underline{c}$, and then it gets the gross profit $1 - t^c - \underline{c}$; City B adopts a good mode of production at the higher cost of $\bar{c}$, and then it get the gross profit $1 - t^c - \bar{c} < 1 - t^c - \underline{c}$. In other words, City A gains a higher economic growth than City B, and causes accidents with higher probability. Against the institutional background of political centralization and economic decentralization, CPC officials' promotion and assessment of their political merits, to a large extent, are based on their performances in boosting economic growth (Bo 1996; Li and Zhou 2005). This indicates that officials in City A will have more opportunities to be promoted than those in City B because the former outperformed the latter in economic construction. This not only explains the concomitance of high growth rate and high frequency of vicious accidents, but also reveals the defining characteristics of the so-called China Model. From this perspective, we can assert that government–firm collusion is the inherent drive of the China Model.

2.4.2 *Collusion-Proof Contract*

Now that collusion brings extra undesirable social costs on the central government, why does it acquiesce to collusion? For this question to be solved, it is necessary to make a comparison between the expected revenues in the case of allowing collusion and those in the case of deterring collusion.

Suppose the central government attempts to prevent collusion, it must design an optimal collusion-proof contract which should satisfy the participation constraint or individual rationality constraint (IR) and the incentive compatibility constraint (IC) of local government and the firm. To be more specific, this optimal collusion-proof contract must satisfy all the following conditions.

First, the firm's individual rationality constraint (AIR) and the local government's individual rationality constraint (SIR) should be both satisfied.

$$\text{(AIR)}\ 1 - t^{cp} - \bar{c} \geq 0 \tag{2.5}$$

$$\text{(SIR)}\ a^{cp} t^{cp} \geq 0 \tag{2.6}$$

Second, the contract should take into consideration the firm's incentive compatibility constraints. When the firm prefers a good mode of production, its expected revenue must be greater than or equal to that which is expected in the case of its choosing a bad one.

$$\text{(AIC)}\ 1 - t^{cp} - \bar{c} \geq 1 - t^{cp} - k\Delta c - p\rho F_A \tag{2.7}$$

Third, the local government's incentive-compatibility constraints (ICC) assure that the expected revenues in the case of noncollusion are greater than or equal to those in the case of collusion.

$$\text{(SIC)}\ a^{cp} t^{cp} \geq a^{cp} t^{cp} + k\Delta c - p\rho F_S \tag{2.8}$$

At last, the firm's limited liability constraints (ALL) and the local government's limited liability constraints (SLL) should also be satisfied. Punitive measures should be less than or equal to their total expected revenues.

$$\text{(ALL)} \quad F_A \leq 1 - \underline{c} - t^{cp} - k\Delta c \tag{2.9}$$

$$\text{(SLL)} \quad F_S \leq a^{cp} t^{cp} + k\Delta c \tag{2.10}$$

According to the assumption, SIR is automatically satisfied. The central government's objective function is to maximize its expected revenue. Therefore, arranging all sorts of constraint conditions, we get the following planning P1:

$$\underset{\{\alpha,t,F_A,F_s\}}{\text{Max}} \left(1-\alpha^{cp}\right)t^{cp}$$

$s.t.$

$$\text{(AIR)}\ 1-\bar{c}-t^{cp}\geq 0$$

$$\text{(AIC)}\ k\Delta c+p\rho F_A\geq \Delta c$$

$$\text{(ICC)}\ p\rho F_S\geq \Delta c$$

$$\text{(ALL)}\ 1-t^{cp}-k\Delta c-\underline{c}\geq F_A$$

$$\text{(SLL)}\ a^{cp}t^{cp}+k\Delta c\geq F_S$$

According to the collusion-proof principles (Tirole 1986), the mechanism satisfying the above planning P1 must be collusion-proof and the principal cannot get more revenues than this. By combining Lagrange function and Kuhn–Tucker theorem, we can come to the Proposition 2.2. (For the proving procedure, see Appendix A.1).

Proposition 2.2 *The optimal collusion-proof contract has the following properties:*

1. *Both AIR and SIR are loose while the other constraint conditions are tight.*
2. *All endogenous variables are determined by the following expressions:*

$$a^{cp}=\frac{(1-p\rho)k\Delta c}{p\rho(1-c)-\Delta c(1-k+p\rho k)} \tag{2.11}$$

$$t^{cp}=1-\underline{c}-k\Delta c-\frac{(1-k)\Delta c}{p\rho}<1-\bar{c} \tag{2.12}$$

$$F_A=\frac{(1-k)\Delta c}{p\rho} \tag{2.13}$$

$$F_S=\frac{k\Delta c}{p\rho} \tag{2.14}$$

3. *The firm's profit is*

$$R = 1 - \bar{c} - t^{cp} = (1-k)\left(\frac{1}{p\rho} - 1\right)\Delta c > 0 \tag{2.15}$$

In Proposition 2.2, conclusion (1) reveals that the expected revenues of both the local government and the firm must be greater than their reservation utility, i.e., they both obtain positive information rent in the case of information asymmetry. This conclusion is different from the moral hazard model where the agent's participation constraint is tight. Since the central government does not know the exact type of the firm's cost, it should give the local government and the firm part of information rent so as to encourage the firm to adopt good mode of production.[21] In the case of deterring collusion, the participant constraints of the local government and the firm can be satisfied even if central government merely gives them zero reward and zero profit. Therefore, a collusion-proof contract reduces the central government's well-being.

Conclusion (2) implies that the optimal tax rate of deterring collusion is t^{cp} and that, under the tax-sharing system, the tax proportion distributed to the local government is α^{cp}. F_A and F_S, respectively, denote the penalties inflicted by the central government on the local government and the firm. The expression of optimal penalty is extremely intuitive: Pecuniary penalty should be equal to expected rent. In other words, under the protection of limited liability, both the central government and the firm gain nothing at most even if the central government knows their collusion.

A further discussion of collusion-proof mechanism is conducive to explaining the implications of some policies.

Corollary 2.1 *The optimal collusion-proof contract has the following comparative statics: (The proving process is omitted here due to its simplicity)*

1. $\frac{\partial \alpha^{cp}}{\partial \rho} < 0$, $\frac{\partial t^{cp}}{\partial \rho} > 0$, $\frac{\partial F_A}{\partial \rho} < 0$, $\frac{\partial F_s}{\partial \rho} < 0$, $\frac{\partial R}{\partial \rho} < 0$;
2. $\frac{\partial \alpha^{cp}}{\partial \Delta c} > 0$, $\frac{\partial t^{cp}}{\partial \Delta c} < 0$, $\frac{\partial F_A}{\partial \Delta c} > 0$, $\frac{\partial F_s}{\partial \Delta c} > 0$, $\frac{\partial R}{\partial \Delta c} > 0$.

The proof of Corollary 2.1 is quite simple and can be omitted here. Conclusion (1) implies the stronger the supervisory ability of the mass media, the less the sharing proportion distributed to the local government, the higher the tax rate, the lighter the penalty, and the lower the information rent given to the firm. So far as the central government is concerned, mass media supervision improves its probability to perceive collusion. Conclusion (2) reveals that the more the quasi-rent, the more the sharing proportion distributed to the government, the lower the tax rate, the heavier the penalty, the more the information rent given to the firm.

2.4.3 *Allowing Collusion Versus Deterring Collusion*

How does the central government decide to allow or deter collusion? Obviously, the key to this essential decision problem lies in making a comparison between the expected revenues, respectively, brought about by equilibrium collusion contract and collusion-proof contract.

Let us consider equilibrium collusion contract first. In the case of allowing collusion, it is not necessary for the central government to urge the local government to supervise the firm's choice of a good mode of production. Thus, the local government's participant constraint is satisfied even if it gets zero reward, and this signifies $\alpha^c = 0$. Now that the firm's reservation utility is also 0, the central government leaves the firm zero profit.[22] We can infer the following:

$$U_A^c = 1 - t^c - \underline{c} = 0 \tag{2.16}$$

$$t^c = 1 - \underline{c} \tag{2.17}$$

It is noted that, in this case, no penalties have been inflicted on both the local government and the firm because the government-firm collusion is "legitimate." As a matter of fact, the central government burdens all extra social cost C even though it gains all surpluses by setting the optimal tax rate. Although the local government negotiates an agreement with the firm without any compromise, it cannot seize any surplus in that the central government has first-mover advantage. Thus, under the equilibrium collusion contract, the central government's expected revenue is:

$$U_P^c = 1 - \underline{c} - pC \tag{2.18}$$

Now let us calculate the central government's expected revenue under the collusion-proof contract. According to proposition 2–2, its expected utility is:

$$
\begin{aligned}
\underline{U}_p^{cp} &= (1-\alpha^{cp})t^{cp} \\
&= \left(1 - \frac{(1-p\rho)k\Delta c}{p\rho(1-\underline{c}) + (1-p\rho)k\Delta c - \Delta c}\right) \times \left(1 - \underline{c} - k\Delta c - \frac{(1-k)\Delta c}{p\rho}\right) \\
&= 1 - \underline{c} - \frac{\Delta c}{p\rho}
\end{aligned} \quad (2.19)
$$

Proposition 2.3 can be obtained from the above comparison between expected revenues in the case of allowing and deterring collusion.

Proposition 2.3 *If and only if $pC \leq \frac{\Delta c}{p\rho}$, the central government allows local government–firm collusion, but not the other way around.*

For Proposition 2.3 to be more intuitive, suppose in an extreme case $p = \rho = 1$, i.e., the choice of a bad mode of production will inevitably incur accidents and they will be undoubtedly disclosed by the mass media, then the central government's maximal expected revenue is $\max \underline{U}^{cp} = 1 - \bar{c}$. In this case, the corresponding critical condition is $C \leq \Delta c.^p$This means that it is worthwhile for the central government to allow collusion so long as the extra social cost is less than the quasi-rent arising from a bad mode of production. When the critical condition is defined as $G \equiv \frac{\Delta c}{p\rho} - pC$, more interesting comparative static conclusions can be drawn.

Corollary 2.2: *If Proposition 2.3 is true, then* $\frac{\partial G}{\partial \Delta c} > 0$, $\frac{\partial G}{\partial p} < 0$, *and* $\frac{\partial G}{\partial \rho} < 0$.

This reveals that ceteris paribus, the higher the quasi-rent, the less probability for accidents to occur, the less probability of the accidents to be disclosed, and the more likely for the government to allow collusion. To conclude, the probability of equilibrium collusion is positively correlated with the collusion-proof cost.

2.4.4 *The Role of Compensation*

The role of compensation is ignored in the above analysis. It is implicitly assumed that the compensation for the direct victims (killed or injured coalminers) of the accidents, if compared to the potential profit of the firm, is negligible. In reality, this assumption is sound. According to statistics, the death toll per million tons of coal during the 6 years between 2000 and 2005 is 4.46. If the average compensation for each victim is CNY 200,000, the total compensation per million tons is approximately

CNY 900,000. The gross margin for production of 1 million tons of coal is at least CNY 100,000,000.[23] Since workers in the high-risk industries such as coalmine enterprises and firecracker factories are predominantly poor peasants whose reservation utility is very low and whose supply far exceeds demand, their participation constraints are theoretically negligible. Now, let us relax this assumption that the compensation paid for accidents is decided by law, what role will it play in the optimal collusion-proof mechanism?

Suppose the probability of vicious accidents per unit of output is still p, and the compensation which the firm should pay is m. A slight modification is made to the optimal collusion-proof contract so as to make sure that the agent pays an additional sum of m in game time sequences (3) and (4) discussed in Sect. 2.3.4 and thus the incentive compatible constraints (AIC) and limited liability constraints (ALL) will be modified accordingly. With the compensation considered, we draw the following proposition.

Proposition 2.4 *On the condition that the compensation is below the threshold* $m = \frac{(1-k)(1-p\rho)\Delta c}{p(1-\rho)}$, *the optimal collusion-proof mechanism is valid, and* $\frac{\partial \alpha^{cp}}{\partial m} < 0$, $\frac{\partial t^{cp}}{\partial m} > 0$, $\frac{\partial F_A}{\partial m} < 0$ (For the proving procedures, see Appendix A.2).

Proposition 2.4 implies that, if the compensation is above the threshold, the optimal collusion-proof contract will be invalidated because the firm, without the lure of information rent, will not choose a good mode of production. If the compensation is below the threshold, the optimal collusion-proof contract is still valid and in this case, compensation plays a certain role in punishing the firm. Compensation has a supervising effect (higher compensation will naturally lower the difficulty degree of supervision) and substitution effect (under the limited liability, higher compensation is likely to reduce the frequency of occurrence of accidents and thus the pecuniary penalty will be accordingly lowered on the whole). It is recommended that, if the compensation is too low, the government should moderately raise statutory compensation so as to make it a positive incentive. Nevertheless, it is advisable that the central government should not excessively rely on compensation for the following three reasons: (1) exorbitant compensation will put the firm at high risk or even into jeopardy because accidents cannot be at any rate eradicated even if a firm is in normal and regular production, and moreover, the execution cost of law should also be taken into account; (2) exorbitant compensation may force coalmine enterprises to connive with local officials and cover up accidents; and (3) concealment of accidents due to higher compensation invalidate punitive measures and make it difficult to get rid of the negative externalities caused by safety accidents. In reality, whether the central government and the mass media can get the first-hand evidence from accident victims or their relatives or not is very important to effectively fight local collusion. Given the coalmine workers' extremely low reservation utility, it is very difficult to stipulate the statutory compensation because many factors should be taken into consideration, such as the incentive of the firm's choice of production mode, the incentive for victims or their relatives' disclosure of accidents, regional income gap, and execution cost of law and regulations.

2.5 A Case Study

This section will be dedicated to the application of the government-firm collusion mechanism to the PX (p-Xylene) project mentioned in 2.1. Two realistic questions will be discussed.

First, why did some local governments doggedly push forward p-Xylene projects despite resistance and pressure from local residents? According to the collusion mechanism, a PX project is a "bad" production mode which, if compared with a "good" production mode such as tourism and banking industries, will bring the local government more GDP and fiscal revenues at a quicker rate. A PX project with total investment of hundreds of billions of Chinese Yuan, if put into normal and regular production, will produce billions of Chinese Yuan of industrial output and further drive up the local GDP aggregate, thus helping the local government build up their administrative performances, a foremost important yardstick of political promotion. In an era during which nearly all local governments give priority to investment attraction,[25] ambitious projects like building PX enterprises were very popular among them.[26] Developing high-investment, high-revenue, and heavy-polluting project is an expression form of local government–firm collusion. According to 2.1, given that the central government allows collusion, the local government has a great motivation to collude with the firm.

Second, why did some projects abort while others were put into operation smoothly or relocated? This question involves the cost and return of collusion.

Dalian is a city oriented with heavy and chemical industries. In 2011, its GDP totaled CNY 615.01 billion, of which primary industry accounts for CNY39.57 billion, secondary industry for CNY 320.42 billion, and tertiary industry for CNY 255.02 billion, with the proportion of them being 6.4:52.1:41.5. In the secondary industry (mainly including manufacturing, construction and mining), the industrial added value amounted to CNY281.6 billion, of which heavy industry accounted for 64% (or CNY 180.61 billion) of the total industrial added value or 30% of the local GDP aggregate. The added value of petrochemical industry was CNY 58.85 billion. For Dalian, the petrochemical industry is the most important industrial component.[27] A PX project would bring much higher expected revenue to the Dalian city government, which therefore, despite the occurrence of massive protest and the PX project official's assault of CCTV (China Central Television) correspondents, runs the risk to willfully push forward this profiteering program. According to proposition 2.3, the central government allows collusion if the expected revenue yielding from collusion exceeds that from noncollusion.

Unlike Dalian, Xiamen's GDP in 2011 totaled CNY 253.58 billion, of which primary, secondary, and tertiary industries, respectively, accounts for CNY2.468 billion, 130.41 billion, and 120.702 billion, with the proportion of them being 1.0:51.4:47.6. In contrast, Xiamen is a tertiary industry-oriented city, whose reputation as a tourist resort is likely to be damaged with the introduction of a PX project because its cost is too high and the prospective revenue will fail to satisfy the city government's expectation. Furthermore, since Xiamen is adjacent to Taiwan, a massive protest may jeopardize social stability and even involve national security. Since the collusion cost is more than the expected revenue, the provincial and city governments decided to relocate the project in the more remote Peninsular Gulei.

The case in Ningbo is totally different. The enterprise responsible for the integration refinery project is a state-run conglomerate of China Petrochem, which is not subject to local jurisdiction. Only 1/50 of over CNY10,000 million of profit and tax would be retained for the Ningbo government, which would have to assume the huge cost of relocating local residents and facing criticism and massive protests. Eventually, after careful consideration, the Ningbo government decided to give up on the PX project.

2.6 Extended Discussions

An obvious question arises: why is it difficult to eliminate coalmine accidents even though the central government repeatedly orders a nationwide crackdown on safety accidents? Major news media direct their spearhead against collusion between government officials and firms. It is true that the rampant government–firm collusion is the culprit of high frequency of vicious accidents.[28] Besides the form of collusion in which the central government acquiesces under some specific circumstance, there are still other reasons involving policy and implementation.

2.6.1 *The Local Government's Lack of Long-Term Expectation*

The above analysis implicitly assumes that the local government in a particular term hopes to get its own expected revenue within a production cycle, i.e., the production cycle is identical to the tenure. In reality, the term of the local governmental leadership and of officials responsible for safety work are determined by many factors with randomicity.[29] Given the local government has a discount factor $\delta \in (0, 1]$, $\delta = 1$ denotes the permanent tenure of the local government, then under the collusion-proof mechanism the local government's incentive compatibility constraint (ICC2) is:

$$\delta \text{ at} \geq \delta(\alpha t + k\Delta c) - p\rho F_s \tag{2.20}$$

Since the constraint condition is tight, the optimal discount factor is:

$$\delta^{cp} = \frac{p\rho F_s}{k\Delta c}$$

If the term of the local governmental leadership is too short, or if there is an unexpected personnel transfer, the incentive compatibility mechanism will be destroyed. Below the optimal value, the less δ, the less the local government's sharing proportion of tax, the more likely for it to collude with the firm and even merge itself with the firm so as to find extra bonanza.[30] The local government may even extort more bribes k from the firm and thus instigate the firm to choose a bad mode of production.

2.6.2 *Imposition of Exorbitant Taxation on the Firm*

In the proof-collusion contract, the optimal tax rate is t^{cp}, which takes into full consideration quasi-rent probability of occurrence of accidents, and the supervising ability of the fourth party. However, the prevalent taxation regime is subject to little change once established by law. Given $t > t^{cp}$, the firm will not be able to undertake the input cost of a good mode of production, but perversely force the firm to choose a bad mode of production.[31] Therefore, it is necessary to deregulate the state-run enterprises so as to let them obtain sufficient information rent.

2.6.3 *Ineffectiveness of Supervision of the Mass Media*

Since the central government does not know the authentic information of the firm's input cost, it relies on the supervision of the mass media for disintegrating the coalition of collusion. Corollary 2.1 signifies the role that the mass media play in reducing the central government's reliance on the local government for the firm's information and containing local officials' abuse of power. However, under the current system, the mass media will be suppressed by the local government if they dare to disclose work safety accidents and other scandals. The disclosed accidents may only be part of a huge iceberg. How to help the mass media defend their legitimate rights in supervising local government–firm collusion deserves our attention.

2.6.4 *Lack of Credibility of Punishment*

The central government's crackdown on collusion lacks credibility for the following two reasons. First, some local officials removed from office may get a post in another city or county or province, rendering administrative punishment on them invalidated. If the credibility of penalty is questionable, the local officials will ignore the central government's relevant regulations.[32] Secondly, when facing the pressure of public opinion due to the occurrence of a serious vicious accident in an enterprise, the central government tends to indiscriminately crack down even all enterprises in the same sector. For the firm, this "time inconsistency" of policy results in a lack of long-term stable expectation to choose a good mode of production (Nie 2005). Improvement of the fourth party's supervising ability is an important means to increase the credibility of the central government's commitment. Fortunately, the safety work department directly under the control of the government has decided to strengthen credibility. For instance, the State Administration of Coalmine Safety has decided to publish the progression of crackdown on illegitimate coalmine enterprises.[33]

In addition to the four factors concerning the implementation of policies, there are still others which interact with each other to make it difficult to deter collusion. By comparison, the annual mineral output of the United States is roughly equal to that of China, but its death toll in 2004 was only 28 (see Table 1.1), much lower than China's 6027. In 1977, the US Congress enacted the *Federal Mine Safety and Health Act*, declaring that "the first priority and concern of all in the coal or other mining industry must be the health and safety of its most precious resource—the miner."[32] It sets forth mandatory safety and health standards and compensation for damage to and loss of properties and victims in the case of accidents. According to this Act, the United Mine Workers of America is responsible for negotiation with the management over wages and safe working conditions; all mine workers must receive a strict training program and safety standards should be strictly enforced. The enforcement of this Act in America validates our above analysis from another perspective.

2.7 Summary

This chapter provides a government–firm collusion model which can simultaneously and endogenously explain the concomitance of high economic growth and high frequency of vicious accidents, thus offering a microbasis interpretation of the China Model. According to our model, under the condition of information asymmetry, local governments collude with firms to choose a bad mode of production for the purpose of bumper fiscal revenue and political promotion. As far as the central government is concerned, if the expected revenue in the case of collusion is more than the expected cost, it will allow collusion; if not, it will deter collusion. In this sense, government–firm collusion is a neutral phrase without prior moral criticism.

The policy implications of this chapter are as follows. If the central government attempts to deter collusion, it should (1) protect mass media's disclosure of illegitimacy of the local government and the firm; (2) minimize the revenue of collusion, which can be realized by subsidizing good modes of production and strictly enforcing laws concerned; and (3) ensure the stability and sustainability of local government's policies and preclude its short-sightedness.

This model can explain a series of mishaps: coalmine accidents, environmental pollution, food safety, housing bubbles, jerry-built projects and smuggling, etc. The forthcoming chapters Three, Four and Five will discuss the applications of this collusion model.

Notes

1. "The Way out for PX Projects", *Caixin Zhoukan* (*Financial Review Weekly*), http://zj.sina.com.cn/news/n/2012-10-29/075229096 (Accessed on November 13, 2012).
2. CNY, denoting Chinese yuan, is the symbol of the Chinese currency renminbi.
3. All the three cases in this section are all collected and arranged from sources such as Wikipedia and Baidupedia, and some influential newspapers and websites.
4. Yang and Nie (2008) makes a critical review on literature concerning market-preserving federalism.
5. However, there is still a controversy about the causality between political promotion and economic achievements. Opper and Brehm (2007) observe that GDP growth rate is not as a weighty criterion for an official to be promoted as some analysts have asserted once he manipulates political network or otherwise called relation (*guanxi*) in Chinese. Tao et al. (2010) makes a logical and empirical criticism against the alleged causality between political promotion and economic achievements.
6. Suppose an economist argues that political promotion results in local economic growth and that malignant competition leads to local protectionism, two problems will arise: what are the cost of political promotion and the revenue of malignant competition respectively? In economics, each option has its own cost and revenue.
7. Among the few of articles and monographs concerning equilibrium collusion are Tirole (1992), Kofman and Lawarree (2007), Lambert-Mogiliansky (1998) and Suzuki (2007).
8. The model in this chapter is principally based on Nie and Li (2006; 2013).
9. Here we temporarily do not distinguish vertical management units (such as customs house) directly affiliated to the central government and from those units under the charge of local governments. We will discuss the impact of the two types of management units on collusion in Chap. 9.
10. For the central government, it is much less costly to use GDP as an indicator of evaluating or measuring officials' political performances. Nevertheless, this can also be interpreted as reflecting public preference.
11. In its strict sense, the reward given to the local government is endogenous. Nevertheless, this linear form of share is the optimal form of remuneration. The basic logical reasoning proceeds as follows: moral hazard is inevitable under the condition that the central government entrusts the local government with a package task (far beyond merely supervising the firm's production activities) but fails to supervise the local government's

activities. To reduce moral hazard, the optimal contract is undoubtedly sharing contract (Bolton and Dewatripont 2005).

12. Though this assumption is different from other contract theories, our analysis is based on real situations. For instance, it is required that a coalmine enterprise, before its operation, should at least get a mining permit, safety production license, coal production license, business registration certificates issued by governmental departments concerned. The local government at the level of county or township is very well-informed of the firm's production mode and accident prevention measures. Even if the local government knows with a certain degree of probability the firm's production technology or techniques, the conclusion is still tenable.
13. Social stability, in its broad sense in this specific context, refers to all incidents that possibly undermine the stability of a regime.
14. Risk neutral plus limited responsibility is equated with risk aversion. This assumption is consistent with Kofman and Lawarree (1993) and Laffont and Martimort (1995). In general, when an accident occurs, the person in charge of work safety is more likely to be given a financial penalty of a certain sum of money instead of a criminal penalty. For instance, according to *Coal Mine Safety Supervision Regulations* (2000), a coal mine supervision organ can give anyone convicted of covering up an accident, sabotaging the accident scene, and obstructing investigation a fine of CNY 30,000 to 150,000. This shows that our assumption is sound and reasonable.
15. According to the *Regulations on Selection and Appointment of Leading Cadres* issued by the Chinese Communist Party, officials in the local government, including party secretariats, provincial governors and mayors, can be elected to only two five-year terms of office. Besides, below the provincial level, any leading cadre with the supreme authority of decision making cannot hold a post in office in his birthplace. A data-based research shows that the average tenure for provincial leaders is 3.3 years. (Li and Zhou 2005).
16. A local government is the supervisor of a firm, so the local government can demand that the firm should accept its decision. Relaxation of this assumption may result in some interesting implications.
17. The multiplication of the value of transfer payment by φ, denoting a coefficient of transaction cost will help an expansion of our research. Faure-Grimaud et al. (2001) analyzed the impact of transaction costs of collusion on prevention of collusion.
18. The central government's optimal strategy is to be discussed in 2.4.3.
19. Technically, we suppose the firm's expected revenue is always positive. Otherwise, the firm will withdraw from this game.

20. In the real world, the local government can always formulate and promulgate investment attracting policy with a certain orientation. For instance, if the local government acquiesces to operation of some polluting enterprises, those green enterprises with high cost input will be dislodged from the local market.
21. Theoretically, whether the moral hazard model or the choice model is chosen, the fundamental principle is that the principal alienates information rent to the agent with the information advantage so that the hazard of information asymmetry can be alleviated and the agent can be honest and diligent.
22. According to Nie and Zhang (2012), it is not necessary for the local government to supervise a firm which chooses a good mode of production; therefore, its reservation utility is the expected profit in the case of good mode of production. In this chapter the assumption that the firm has no right to choose modes of production is a form of simplification, which will have little or no effect on the major results.
23. The figure is calculated according to the data sorted out by the author of this book through www.coalworld.net.cn and www.chinanews.com.
24. A safety accident (such as an explosion) incurs losses of lives and properties; in addition, it pollutes the local environment. If the direct victims (workers) do not provide evidence of the accident, the indirect victims (local residents) find it more difficult to protect their own legal interests.
25. In China, local governments at all administrative levels were so obsessed with "investment attraction" that it became a defining characteristic over the past three decades. There was an extreme governmental slogan: WHOEVER OBSTRUCTS INVESTMENT ATTRACTION SPITES THE JIAOHE PEOPLE. Jiaohe is a county in the east of Jilin Province, Northeast China.
26. In 2011, I conducted a survey concerning industrial zones in provinces of Jiangsu and Anhui, where the chief officials told me that the local governments in many regions set a higher threshold for enterprises which hope to enter the industrial zone, demanding that a minimum of fixed assets investment should be CNY 2.5 million per *mu* (a Chinese measurement unit which is approximately 667 square meters), sales volume CNY 5 million per *mu* and tax revenue CNY 0.15 million per *mu*.
27. For particulars, see the Dalian government's official website http://www.dl.gov/dlgk/zxsq (Accessed April 16, 2012).
28. On March 14, 2005, a terrible gas explosion broke out in a coalmine pit in Qitaihe, Heilongjiang, Northeast China, killing 18 workers. Later, the investigation shows that the coalminer Peng Guocai was vice-director of the Administration of Work Safety, Taoshan District, Qitaihe.
29. According to the current system, a higher local government's arrangement of transfer, promotion and removal of officials in a lower local

government is not strictly identical to the tenure of the local collective leadership. As a matter of fact, the standing committee of the local people's congress has the authority to approve the appointment and removal of any local official at any time without having to wait until the expiration of the five-year-long office term.

30. To some extent, this is perhaps the reason for a local government's connivance at illegal operation of dangerous small coal mines, diggings and fireworks workshops because a local government especially below city level cannot find any way to get any revenue from a state-run large- and medium-sized enterprise but can get profit from small plants and factories within the area under its own jurisdiction. Due to unpredictable tenure, a local official may be reluctant to encourage an enterprise to adopt a good mode of production even if they become shareholders of an enterprise.
31. According to our survey, there are two reasons for state-run coalmine enterprises' reluctance to choose a good mode of production: (1) lack of necessary capital accumulation—due to the sluggishness of the coal industry prior to 1989—to renovate their production and safety facilities; (2) the policy of "compensation for the shortage of electricity with coal" due to the increasing demand for electricity with the acceleration of national industry. According to a report on February 9, 2005 carried in a newspaper *Xinjingbao* or *The Beijing News*, approximately CNY68,900 million would be needed to renovate coalmine safety facilities throughout the country.
32. For instance, on August 22, 2005, the General Office of the State Council issued *The Urgent Notice Concerning Crackdown on Coalmine Enterprises without Safety Production Conditions*, which adjured that functionaries of state organs and state-run enterprises should withdraw their investment in coalmine enterprises. However, no functionaries voluntarily did so by the deadline and many local governments even extended the deadline.
33. See the speech addressed by Li Yizhong, director of the State Administration of Work Safety, in a safety work videoconference on April 3, 2003 (http://www.chinasafety.gov.cn/zhengwugongkai/2006-04/04. Accessed June 18, 2003).
34. "U.S. Mine Safety and Health Administration: 25 Years of Success," (U.S. Department of Labor, 2003 http://www.msha.gov/MSHAInfo/25Years/MSHA%2025%20Years.pdf (Accessed June 18, 2011).

References

Bai, Chong-En, Yingjuan Du, Zhigang Tao, Sarah Tong. "Local Protectionism and Regional Specialization: Evidence from China's Industries". *Journal of International Economics*. 2004, 63 (2): 397–417.

Bardhan, Prana & Dilip Mookherjee. "Capture and Governance at Local and National Levels". *American Economic Review*. 2000, 90 (2): 135–139.

Bo, Zhiyue. "Economic Performance and Political Mobility: Chinese Provincial Leaders". *Journal of Contemporary China*. 1996, 5 (12): 135–154.

Bolton, Patric and Mathias Dewatripont. *Contract Theory*. Cambridge: MIT Press, 2005.

Brandt, Loren and Thomas Rawski. *China's Great Economic Transformation*. UK: Cambridge University Press. 2008.

Cai, Hongbin, and Daniel Triesman. "State Corroding Federalism". *Journal of Public Economics*. 2004, 88: 819–843.

Chen, Ye, Hongbin Li and Li-An Zhou. "Relative Performance Evaluation and the Turnover of Provincial Leaders in China." *Economic Letters*. 2005, 88: 421–425.

Cheung, Steven. *The Economic System of China*. Hong Kong: Arcadia Press, 2008.

Chow, Gregory and Kui-Wai Li. "China's Economic Growth: 1952–2010". *Economic Development and Cultural Change*. 2001, 51 (1): 247–256.

Faure-Grimaud, Antoine. et al. "Transaction Costs of Collusion and Organizational Design". USC Center for Law, Economics & Organization Research Paper No. C01–17, 2001.

Faure-Grimaud, Antoine, et al. "Collusion, Delegation, and Supervision with Soft Information". *Review of Economic Studies*. 2003, 70 (2): 253–280.

Gordon, Roger. "An Optimal Taxation Approach to Fiscal Federalism". *Quarterly Journal of Economics*. 1983, 98 (4): 567–586.

Jin, Hehui, Yingyi Qian and Barry Weingast. "Regional Decentralization and Fiscal Incentives: Federalism, Chinese Style." *Journal of Public Economics*. 2005, 89 (9–10): 1719–1742.

Kofman, Fred, and Jacques Lawarree. "Collusion in Hierarchical Agency". *Econometrica*. 1993, 61: 629–656.

Kofman, Fred and Jacques Lawarree. "On the Optimality of Allowing Collusion". *Journal of Public Economics*. 1996, 61 (3): 383–407.

Laffont, Jean-Jacques and David Martimort. "Collusion and Delegation". *Rand Journal of Economics*. 1995, 29 (2): 280–305.

Lambert-Mogiliansky, Ariane. "On the Optimality of Illegal Collusion in Contracts". *Review of Economic Design*. 1998, 3: 303–328.

Li, Hongbin and Li-An Zhou. "Political Turnover and Economic Performance: The Incentive Role of Personnel Control in China." *Journal of Public Economics*. 2005, 89: 1743–1762.

Lin, Justin Yifu and Zhiqiang Liu. "Fiscal Decentralization and Economic Growth in China". *Economic Development and Cultural Change*. 2000, 49 (1): 1–21.

Maskin, Eric, Yingyi Qian and Chenggang Xu. "Incentives, Scale Economies and Organization Forms". *Review of Economic Studies*. 2000, 67 (2): 359–378.

Nie, Huihua. "A New Perspective of Reducing Safety Accidents". *Globe*. 2005 (1): 11–12. 聂辉华. 减少生产安全事故要有新思维. 环球. 2005 (1): 11–12.

Nie, Huihua. "The Impact of Cancellation of Agricultural Tax on Township Governments' Behaviors: An Analytical Framework of Entrusted Agency". *The Journal of World Economy*. 2006, 8: 71–78. 聂辉华, 取消农业税对乡镇政府不同行为的影响: 一个多任务委托代理分析框架. 世界经济, 2006, 8: 71–78.

Nie, Huihua and Jinbo Li. "Collusion and Economic Growth". *China Economic Quarterly*. 2006, 6 (1): 75–90. 聂辉华, 李金波. 政企合谋与经济发展. 经济学(季刊). 2006, 6(1): 75–90.

Nie, Huihua and Jinbo Li. "Collusion and Growth: A New Perspective on China Model". *Economic and Political Studies*. 2013, 1 (2): 18–39.

Oates, Wallace and Robert Schwab. "Economic Competition Among Jurisdictions: Efficiency Enhancing or Distortion Inducing". *Journal of Public Economics*. 1988, 35: 333–354.

Oi, Jean. (1992) "Fiscal Reform and the Economic Foundations of Local State Corporatism in China". *World Politics*. 1992, 45 (1): 99–126.

Opper Sonja, Stefan Brehm. "Economic Performance and Networks: Political Careers in China's M-Form State". Working paper, 2007.

Qian, Yingyi and Barry R. Weingast. "China's Transition to Markets: Market-Preserving Federalism, Chinese Style". *Journal of Policy Reform*. 1996, 1 (2): 149–185.

Qian, Yingyi and Chenggang Xu. "Why China's Economic Reforms Differ: the M-form Hierarchy and Entry/expansion of the Non-state Sector". *Economic Transition*. 1993, 1 (2): 135–170.

Shen, Liren and Yuanchen Dai. The Formation of China's Fragmented Economy and its Disadvantages and Roots. *Economic Research Journal*. 1990, (3): 12–19. 沈立人, 戴园晨. 我国诸侯经济的形成及其弊端和根源. 经济研究. 1990, (3): 12–19.

Suzuki, Yutaka. "Collusion in Organizations and Management of Conflicts through Job Design and Authority Delegation". *Journal of Economic Research*. 2007, 12 (2): 203–241.

Tao, Ran, et al. "Can Economic Growth Benefit Political Promotion? A Logic Challenge Against Promotion Tournament Theory and a Provincial-Level Evaluation". *Management World*. 2010, (12): 13–26. 陶然等, 经济增长能带来晋升吗?—对晋升锦标赛理论的逻辑挑战与省级实证评估. 管理世界. 2012, (12): 13–26.

Tirole, Jean. "Collusion and Theory of Organizations", in Jean-Jacques Laffont, eds. *Advances in Economic Theory: Proceedings of the Sixth World Congress of the Econometric Society*. Cambridge: Cambridge University Press. 1992.

Tirole, Jean. "Hierarchies and Bureaucracies: On the Role of Collusion in Organizations". *Journal of Law, Economics, and Organization*. 1986, 2 (2): 181–214.

Weingast, R. Barry. "The Economic Role of Political Institutions: Market-Preserving Federalism and Economic Development". *Journal of Law, Economics and Organization*. 1995, 1 (1): 1–31.

Yang, Qijing, and Huihua Nie. "Market-Preserving Federalism and Criticism Thereon". *Economic Research Journal*. 2008, (3): 99–114. 杨其静, 聂辉华. 保护市场的联邦主义及其批判. 经济研究, 2008, (3): 99–114.

Yao, Yang. "Is There Really a China Model?" *Reprinted Journal Materials of Renmin University, "Socialism Economic Theory and Practice"*. 2008 (10). 姚洋. 是否存在一个中国模式?人大复印报刊资料 "社会主义经济理论与实践", 2008 (10).

Young, Alwayn. "The Razor's Edge: Distortions and Incremental Reform in the People's Republic of China". *Quarterly Journal of Economics*. 2000, 115 (4): 1091–1135.

Zhang, Weiying and Shuhe Li. "Regional Competitions and Privatization of State-Run Enterprises". *Economic Research Journal*. 1998, (12:) 13–22. 张维迎, 栗树和. 地区间竞争与中国国有企业的民营化. 经济研究. 1998, (12:) 13–22.

Zhang, Yan. Yongqin Wang, Yuan Zhang, Zhaochen, and Ming Lu. "China's Development Model: the Costs and Benefits of the Decentralization Approach to Transition". *China Economist*. 2007, 4: 22–33.

Zhou, Li-An. "Incentives and Cooperation Among Governmental Officials in Promotion Game: An Analysis of the Underlying Reason for Localism and Repeated Constructions in China." *Economic Research Journal*. 2004, (6): 33–40. 周黎安. 晋升博弈中政府官员的激励与合作——兼论我国地方保护主义和重复建设问题长期存在的原因. 经济研究. 2004, (6): 33–40.

CHAPTER 3

Collusion and Coalmine Accidents

By analyzing the data collected from the Chinese coal industry, this chapter mainly validates that the primary cause of coalmine accidents is collusion and provides empirical evidence for testing collusion in the hope of helping those interested students apply this approach to other fields. Section 3.1 makes a survey of coalmine accidents in China and then a literature review in this respect. Section 3.2 gives an account of the institutional background of coalmine safety supervision. Section 3.3 discusses the econometric approaches to collusion in the coal industry. Section 3.4 is an analysis of the results obtained by using the provincial-level data in the coal industry. Section 3.5 provides a robustness testing. Section 3.6 is a summary.

3.1 High Frequency of Coalmine Accidents

China is the largest coal producer in the world, but meanwhile, its mortality rate is also higher than that of any other coal producer of the world. In 2007, China produced 2.523 million tons of coal, accounting for 41.4% of the world total; America's coal output is second only to China, accounting for 18.7%.[1] However, in that year, the death toll arising from coalmine accidents in China was 3958, but only 34 in America.[2] If calculated in terms of per million tons of coal, China's mortality rate between 1988 and 2002 was not only far more than that of advance countries such as America, Japan, and Germany but also dozens of times more than that of developing countries such as India and

H. Nie, *Collusion, Local Governments and Development in China*,
DOI 10.1007/978-981-10-5059-6_3

Table 3.1 Mortality rates of coalmine workers in major countries

Year	*China*	*US*	*Japan*	*South Africa*	*India*	*Germany*
1950		1.27	19.9		9.3	5.13
1955		0.94	15.6			4.31
1960		0.83	11.72		4.81	2.46
1965		0.54	1.75			
1970		0.47	4.26		3.05	1.08
1975		0.26	3.57			0.78
1980	8.17	0.17	1.22	0.89	1.33	
1981		0.21	7		1.26	
1982		0.16	1.36	0.99	1.28	
1983		0.09	1	1.49	1.24	
1984		0.16	5.99	0.68	1.09	
1985		0.08	4.9		1.21	
1986		0.09	0.93		1.11	0.48
1987		0.08	0.54	0.54	0.85	0.39
1988	6.78	0.06	0.64	0.3	0.93	0.34
1989	6.64	0.07	0.54	0.3	0.84	0.41
1990	6.1	0.07	0.68	0.28	0.54	0.26
1991	5.21	0.07	0.24	0.24	0.63	0.32
1992	4.65	0.06	0.12	0.23	0.69	0.45
1993	4.78	0.06	0.49	0.25	1.18	0.22
1994	5.15	0.05	0.28		0.75	0.29
1995	5.03	0.05	0.15		0.54	0.26
1996	4.67	0.04	0.32		0.48	0.25
1997	5.1	0.03	0	0.6	0.54	0.19
1998	5.02	0.03	0.47		0.47	0.19
1999	5.3	0.03	0.47		0.45	
2000	5.86			0.13	0.46	
2001	5.13		0	0.08		
2002	4.64	0.03				

African countries (see Table 3.1). Such a high mortality rate cannot be simply ascribed to the level of economic development when technology and capital moves freely across the globe and when China is in the leading position in economic development. Even if when we take into consideration the fact that China's real GDP per capita between 1992 and 2002 was roughly the same as America's real GDP per capita between 1957 and 1967 and Japan's between 1965 and 1975 (Heston et al. 2006), China's mortality rate among underground coal workers between 1992 and 2002 was still much higher than that in America between 1957 and 1967 and that in Japan between 1965 and 1975. To prevent coalmine accidents, the Chinese government promulgated a series of laws and regulations, such as *Law of the People's Republic of China on Safety in Mines* (1992), *Law of the People's Republic of China on the Coal Industry* (1995), *Regulations of Coal Mine Safety Supervision* (2000), *Work Safety Law* (2002), and *Licensing Regulations of Work Safety* (2004) and meanwhile it readjusted coalmine safety supervision system. In 2000, the Chinese government founded the Administration of Coal Mine Safety, which was later upgraded as a department affiliated to the State Council in 2003 and as a state administration in 2005. However, the mortality rate still remained as high as 1.182 per million tons of coal production in 2008. Coalmine safety management is still an arduous and tremendous task.

Both the academic community and the mass media have attempted to explain the high mortality rate of coalmine workers.

Zhong (2007) asserted that it was the medium- and small-sized coal enterprises' insufficiency of certificates, unclearly defined property rights, and negligence of supervision that resulted in 70% of death toll between 1994 and 2005. However, when the government closed more than 50, 000 medium- and small-sized coal enterprises and reduced one-third of coal production between 1997 and 2001 (Wang 2006), the death toll in this period was still over 5.0 per million tons of coal production and did not drop noticeably (see Table 3.1). Obviously, this assertion failed to provide a satisfactory explanation.

It was also believed that insufficient capital input in the coal mine industry contributed to high mortality rate. According to Tang and Guo (2006), it was the heavy debt accumulated by both big state-run (CNY 68.9 billion) and small township coal enterprises (CNY 200 billion) that led to the shortage of money which otherwise should be inputted to improve safety facilities and gas delivery system and to recruit coalmine safety technicians. Granted, shortage of money is one of the reasons for high mortality rate. Ironically, why do local governments and firms not plough back a certain percentage of profits out of this profiteering industry to improve safety conditions? The question does not really lie in the introduction of first-rate safety equipment, but in the incentive mechanism behind their behaviors.

Some scholars (Cao et al. 2007) asserted that it was coal workers' negligence and poorer education background and lower labor skill that led to higher mortality rate. This viewpoint proves to be problematic too. Indeed, the majority of the underground coal workers are predominantly peasants who have received only primary and junior high school education at most, but it is hard to explain why there is a large difference of mortality rates in different provinces and in different years. By screening the data provided by the State Administration of Work Safety, we find that pernicious coalmine accidents frequented some regions. For example, during the 8 years between 2001 and 2008, Dongliao County in Jilin Province and Qitaihe City in Heilongjiang Province were hit by 26 and 22 coalmine accidents, respectively. Can "negligence of safety" effectively account for repeated occurrences of such calamities in the same region?

Next comes the government-firm collusion model, which holds that, under the condition of information asymmetry, the local government colludes with coal enterprises through reducing safety capital input and deregulating safety supervision for the purpose of higher economic growth, political promotion, and fiscal decentralization (Nie and Li 2006). Once coal production is in positive correlation with the local government's administrative performance, collusion will inevitably result in the curtailment of capital input into safety facilities. Therefore, this model is the institutional factor of coalmine accidents. The difficulties are: how to identify collusion between local governments and coal enterprises and how to empirically test the collusion theory based on information asymmetry and repeated game. These are the questions to be addressed in this chapter.

In order to test several major assumptions that attempt to interpret coalmine accidents, this chapter takes a sample of the coalmine accidents that occurred among key state-run coal enterprises at the provincial level. Especially, three critical variables about the vice provincial governor in charge of safety work are used to measure the level of collusion: (1) *Native*, whether the vice governor is a native or not; (2) *Tenure*5, whether the vice governor is in the 5th year of his tenure; and (3)*Age*50, whether the vice governor is aged over 50 or not when elected as a vice-governor. The results show that

i. during 1995 and 1997, when the key state-run coal enterprises were under the control of the central government, insufficient input in work safety is the main factor of coalmine fatalities;
ii. during 1998 and 2002, when the key state-run coal enterprises were delegated to the lower provincial government, collusion had a great impact upon coalmine fatality rate;

iii. during 2003 and 2005, when the State Administration of Work Safety was upgraded as a department affiliated to the State Council, the coalmine fatality rate is drastically reduced.

The results pass robustness test. For the first time, the data collected from the provincial level have been used to test the impact of collusion on coalmine accidents.

3.2 The Evolution of China's Coalmine Safety Supervision System

Of the sources of energy supply in China, coal accounts for 70%, far above the share of crude oil, natural gas, and hydraulic power. Coal enterprises can be divided into three categories: state-owned key enterprises, state-owned local enterprises, and township enterprises. Before 2002, there still existed another category of state-owned small coal pits.[3] According to the statistics provided by RESSET/DB of Tsinghua University in 2008, state-owned key coal enterprises produced half of the national total coal output, state-owned local enterprises one-sixth, and the township enterprises one-third. However, the death toll of workers in township enterprises accounted for three-fourth, and in the state-owned key coal enterprises, the figure was only half of that of state-owned local enterprises (Wang 2006). Like state-owned enterprises in other industries, state-owned key enterprises had long been in the red in the 1980s and the 1990s because they assume too many social and policy responsibilities, and meanwhile, face unfair competition from township enterprises. In 1998, the State Council delegated 94 state-owned key enterprises directly under the control of the former State Department of Coal and their affiliated 174 enterprises and institutions, together with CNY 237.9 billion assets, CNY 100 billion claims and debts, 3.2 million employees, and 1.33 million retired workers to the lower provincial level. Through this readjustment, total tax revenue should be submitted to the provincial fiscal authorities and the profits should be left for the enterprises themselves.[4] From then on, the State Administration of Work Safety have been responsible for inspecting work safety of state-owned key enterprises, and the provincial governments have been responsible for urging rectification and improvement of work safety facilities.

China witnessed the changes of its coalmine safety supervision system over the past six decades. After the founding of the People's Republic of China in 1949, the State Fuel Department (which lasted for less than 1 year) established a safety supervision administration. By 1980s and 1990s, work safety was put under the charge of the State Coal Department or the State Energy Department. In the meantime, work safety supervision departments were also set up in succession at different levels of local governments of provinces, prefectures, cities, counties, and down to the townships. In 2000, the State Council decided to establish the State Administration of Coal Mine Safety (SACMS), merging office work with the State Bureau of Coal Industry (SBCI) and affiliated to the State Economic and Trade Commission (SETC). In 2001, the SBCI was terminated and the SACMS merged office work with the newly established State Administration of Work Safety (SAWS). The SACMS, for the first time, became independent of coal production administration system. Despite the fact that the SACMS was only responsible for coalmine safety instead for production and that it had established a vertical national administration system at different levels from provinces down to townships, and its affiliates were still under the control of the SETC. In 2003, the SAWS was upgraded as a vice-ministerial-level department affiliated to the State Council and came into its own as an independent coalmine safety administration which was legally free of intervention from local governments. However, since coalmine safety work closely concerns the local government's political performances and the enterprise's interests, conflict of interests and overlapping functions existed between the SAWS and local coal bureaus. In view of this, the State Council promulgated Regulation No. 79 in 2004, clearly defining the responsibility system of "the central government monitoring, the local government supervising, and the enterprise producing." By 2005, the SAWS was further upgraded as a ministerial-level department.

3.3 Testing Approaches to Government-Firm Collusion

According to the classical collusion model (Tirole 1986; Laffont 2000), a principal offers a supervisor and an agent a "take-it-or-leave-it" grand contract and the supervisor and the agent sign a private contract between them. When the agent's private information is known to the supervisor, the two parties may collude and try to maximize their interests by concealing information and behaviors. The principal may crumble this collusion through incentive or bureaucratic policies. If the collusion cost is more than the expected revenue, then collusion will become an equilibrium. Nie and Li (2006, 2013), basing their research on the principal-supervisor-agent analytical framework, construct a government–firm collusion model and discuss the conditions of equilibrium collusion. Government–firm collusion is different from official–businessman collaboration in that the former accentuates mutually beneficial behaviors between the government and the firm such as relaxation of vetting processes of issuing mining licenses, safety works licenses, coal production certificates, miner qualification certificates and business licenses; negligence of day-to-day safety surveillance; reduction of capital input for safety work; concealing accidents and concomitant loss of lives and property; connivance at local officials' equity investment in coal enterprises, etc. The former is an institutional phenomenon while the latter is an individual behavior; the former may be lawful while the latter is usually illicit. Combating bribery and preventing officials from becoming shareholders of a company may reduce, to some degree, the possibility of official–businessman collaboration but may not contain collusion.

The government–firm collusion model provides a new perspective on the explanation of high frequency of coalmine accidents. Empirically, it is difficult to collect direct evidence for collusion because much information is hidden. However, it is still possible to find some explicit characteristics that, with a certain probability, may lead to collusion. The collusion theory assumes that the private contract between the supervisor and the agent is based on repeated game, i.e., the supervisor–agent alliance is dependent on reputation, retaliation, and reciprocality (Tirole 1992). Therefore, we predict that collusion between local governments and coal enterprises will happen with a high probability when their relation depends on reputation, retaliation, and reciprocality, which can function as the indirect evidence for collusion.

As for the state-owned key coal enterprises, their interests have been closely associated with provincial governments since 1998 when the central government delegated them to local governments. The enterprises depend on provincial governments for favorable infrastructures and policies, and provincial governments depend on enterprises for tax revenue, economic growth, and political interests (promotion). Customarily, this type of enterprise is under the charge of a vice provincial governor (or a vice president of an autonomous region, or a vice mayor of a municipality) who is also responsible for other types of industries and GDP growth; so, the identity of the vice provincial governor in charge of economy is the key to understanding of collusion.[5] Suppose collusion rests with the interests of the provincial (or municipal) government represented by this vice governor (or mayor) on the one hand and with his personal interests and political promotion on the other. If the two kinds of interests coincide with each other, the vice governor is prone to strengthen collusion; if not, collusion is likely to be weakened. Seen from the perspective of repeated game, a native vice governor is more likely to collude with enterprises than an outlander (one vice governor transferred from any other province). In addition, the tenure of the vice governor is also related with collusion. In the 5-year tenure, it takes some time for him to be familiar with the specific situations and then build relations and therefore collusion usually happens in the fifth year of his vice governorship. According to law, vice-ministerial officials shall retire at the age of 60 and ministerial officials at 65. Usually, if the vice governor has not reached his retirement age after two tenures, he is likely to be promoted to the ministerial level. Hence, ceteris paribus, we expect that a vice governor aged under 50 is not likely to collude with enterprises.[6] Now, we design three dummy variables: *Native*, *Tenure* 5, and *Age* 50. We assign the value 1 to each variable if it suffices for the condition, and 0 if not, and observe how these variables affect the death rate of coalmine accidents. Moreover, the vice governor's working experience may also affect the probability of collusion, so the variable *Exper* is also taken into consideration. In robustness testing, we add another variable *Tenurek* so as to easily comprehend the function of tenure. In order to observe whether the identity of the governor will have an effect on coalmine accidents or not, we add the variable *Native_g* (denoting whether the governor is a native or not). It should be noted that, even though the above variables are mainly the individual characteristics of the vice governor, they reflect not merely the relationship that binds the local government and coal enterprises but also the inherent conflict and incentive compatibility between the local government and the individual vice governor. For instance, if an outgoing vice governor propels collusion between the provincial government and an enterprise, he will receive a form of an award from the government and, in addition, a handsome sum of bribes from that enterprise. If the vice governor refuses to do so, the local government will promote another official to replace him simply because he could not bring any physical interests.[7] The ways whereby we find evidence for collusion is different from those in currently prevailing politically related literature by Fisman (2001) and Faccio (2006), who attempted to find the explicit correlation (such as the politician's being chosen as director of the board or advisor) to judge how politicians help firms.

In testing the assumption of "insufficiency of capital input," we add to the explanatory variables *Fixasset* (denoting fix assets of mining industry), a variable reflecting safety input in technology and facilities. Besides, we take into consideration the variable *Wage_per* in testing the assumption of "coal worker's poor education" so as to reflect the enterprise's capital input in manpower and human resources. In view of different levels of economic development in different provinces, we take *Rgdp_per* (denoting real GDP per capita) as a controlled variable. In order to reflect cut-throat competition from state-owned local enterprises under the charge of a city or a county, we calculate *Coal_rate* (denoting the percentage of state-owned key enterprises' coal production in the total coal output of the province), a variable which can reflect their market position and can also be used to partially test the assumption that medium- and small-sized coal enterprises are the main source of death toll. Wang (2006) argued that light punishment was one of the main reasons for the difficulty to contain coalmine accidents.[8] The variable *Law_rate* (denoting the percentage of the expenditure on judicial departments in the whole provincial financial budget) indicates law enforcement. Taking a natural logarithm for each of the three variables of *Fixasset, Wage_per,* and *Rgdp_per* so as to smooth data, we establish an econometric equation of panel data as follows:

$$\begin{aligned} Deathrate_{it} = \alpha &+ \beta_1 Native_{it} + \beta_2 Teunure5_{it} + \beta_3 Age50_{it} \\ &+ \beta_4 Exper_{it} + \beta_5 LnFixasset_{it} + \beta_6 LnWage\_per_{it} \\ &+ \beta_7 LnRgdp\_per_{it} + \beta_8 Coal\_rate_{it} \\ &+ \beta_9 Law\_rate_{it} + \varepsilon_{it} \end{aligned}$$

3.4 Results of Econometric Analysis

3.4.1 *Sources of Data*

From *Coal Industry Year Book of China*, we collected the data about coal production, death toll, and death rate of the state-owned key coal enterprises in 22 provinces, municipalities, and autonomous regions (excluding Tianjin, Shanghai, Hainan, and Tibet where no coal is produced, and Fujian, Hubei, Guangdong, Guangxi, and Qinghai where there are no state-owned key coal enterprises) between 1995 and 2005. In addition, we do not take into account the data of Xinjiang (or Sinkiang) and Inner Mongolia because in these two autonomous regions, there are too many opencast coal pits where the death toll and death rate are much lower than those in underground coal pits of other provinces. Thus, the death rates of 20 provinces, autonomous regions, and municipalities are taken as the analytical sample and they are treated as dependent variables. There are two reasons for us to choose the data of state-owned key coal enterprises. First, compared with state-owned local and township coal enterprises, these enterprises are less likely to cover up accidents and therefore, the data are much more accurate. Second, state-owned key coal enterprises have undergone the institutional shift from central control to local administration, which helps us understand how the collusion between the local government and coal enterprises affect death rate. In the dependent variables, information (such as birthplace, age, tenure and working experience) of the vice governor is collected from provincial governmental communiqués and officials' résumés in people.cn (a news platform created by China's most influential *People's Daily*), fixed assets from *China Statistical Year Book*, wage per capita of coal workers, and GDP per capita (readjusted as real GDP per capita in terms of GDP deflator) from stats.gov.cn (official website of the China National Bureau of Statistics), coal output of state-owned key and local enterprises from *Coal Industry Year Book of China*, the expenditure on the judicial system from CEInet Statistics Database. The coal price index in the robustness test is from *China Price Year Book*. In order to solve the endogenous problem, we take the succession after the former vice governor's abnormal dismissal from office (excluding transfer to other positions with the same vice ministerial level, promotion, and retirement), i.e., the present vice governor is transferred or promoted to the current position because in the tenure of the former vice governor a certain pernicious accident occurred. In our sample, 11 vice governors hold office in such an endogenous way, accounting for 20% of the observed values. This indicates that the central government is keenly alert to coalmine accidents and inflicts severe punishment on those vice governors conniving at collusion.

3.4.2 Data Description

The collated and rearranged sample contains 172 observed values of 20 provinces, autonomous regions, and municipalities over a span of 11 years between 1995 and 2005. From Table 3.2, we know that in

Table 3.2 A descriptive statistics of major variables of coalmining (1995–2005)

Codes of Variables	*Unit*	*Averaged value*	*Standard deviation*	*Minimal value*	*Maximal value*	*Observed number*
Deathrate	Per million tons	2.98	6.05	0	66.96	166
Native	Dummy variable	0.44	0.50	0	1	172
Tenure 5	Dummy variable	0.11	0.31	0	1	172
Exper	Dummy variable	0.13	0.33	0	1	172
Age 50	Dummy variable	0.56	0.50	0	1	172
Ln Fixasset	Ln million CNY	209	1.99	−3.91	5.80	168
Ln Wage_per	Ln CNY	9.19	0.44	8.39	10.16	170
Ln Rgdp_per	Ln CNY	8.81	0.59	7.51	10.24	171
Law_rate	Percent	0.83	0.30	0.25	1.67	170
Coal_rate	Ratio	0.49	0.25	0.02	1	168

those years the averaged value of *Deathrate* in the allegedly safest state-owned key coalmine enterprises was 2.98 and the peak value was 66.96. Figure 3.1 reveals the changes of death rate from 1995 to 2005. The death rate hit a record high of 66.96 in 1998, when, according to what is discussed in 3.2, state-owned key coal enterprises were delegated from the central government to provincial governments for management. The State Administration of Coal Mine Safety was established in 2000, so death rate began to decline since then. This indicates that the characteristics of the sample coincided with our intuition. In the sample, 44% of vice governors are natives, and half of them are under age 50 when they hold vice governorship, and only 13% of them have worked for departments or agencies concerning safety work. In addition, more than 70% of them hold their position less than 5 years and only 3% hold office for 10 years (two tenures), and 10% are over age 55 when first elected as vice governor.

3.4.3 *Main Results*

During our sample time (1995–2005), the coalmine safety supervision system of China has undergone three phases. Between 1995 and 1997,

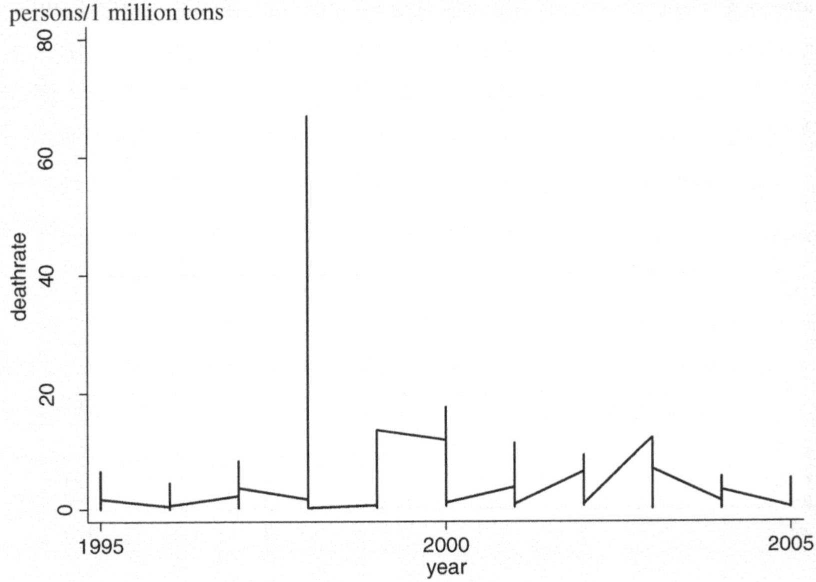

Fig. 3.1 Death rates of state-owned key coal enterprises (1995–2005)

since safety supervision of state-operated key coal enterprises was under the charge of the State Coal Department and had nothing to do with local governments both economically and administratively, we assume that collusion in that period of time had no obvious effect on the mortality rate of state-run key coal enterprises. Between 1998 and 2002, state-operated key local enterprises were delegated to provincial level for administration, and therefore there existed a close relationship between them. Although the State Administration of Coal Mine Safety (SACMS) was established in 2000, its local branches in each province, prefecture, and county were still under the manipulation of local governments. Hence, we assumed that collusion had a significant impact upon mortality rate. Between 2003 and 2005, the State Administration of Work Safety (merging office work with SACMS) was upgraded as a department affiliated to the State Council and came into its own as an independent coalmine safety administration which was legally free of intervention from local governments. Therefore, we assume that collusion had a slight

Table 3.3 Contributors to death rates of state-owned key coal enterprises (1)

Dependent variable: Death rate					
Phases	*1998–2002*			*1995–1997*	*2003–2005*
Model	*(1) RE*	*(2) FE*	*(3) RE*	*(4) RE*	*(5) RE*
Native	4.688**		4.806**	0.179	0.381
	(0.033)		(0.044)	(0.772)	(0.582)
Tenure 5	5.644**		5.526*	0.485	1.341
	(0.050)		(0.060)	(0.571)	(0.205)
Exper	1.013		0.704		
	(0.819)		(0.876)		
Age 50		13.627***	0.398		0.870
		(0.002)	(0.871)		(0.161)
LnFixasset	−0.884	0.604	−0.913	−0.497**	−0.100
	(0.144)	(0.653)	(0.145)	(0.052)	(0.635)
Ln Wage_per	−4.394	−7.222	−4.209	0.883	−2.301
	(0.415)	(0.659)	(0.455)	(0.588)	(0.192)
LnRgdp_per	1.132	35.414	0.947		
	(0.693)	(0.257)	(0.765)		
Law_rate	−2.207	−31.630**	−2.206	0.833	1.280
	(0.588)	(0.014)	(0.601)	(0.588)	(0.402)
Coal_rate	−3.285	−27.590	−3.708	−2.004	−5.637***
	(0.573)	(0.418)	(0.595)	(0.275)	(0.001)
Constant term	36.589	−207.287	36.291	12.207	14.896
	(0.361)	(0.137)	(0.374)	(0.393)	(0.273)
Hausman value	0.523	0.001	0.817	0.611	0.092
R^2	0.223	0.310	0.223	0.293	0.547
Observations	71	71	71	39	51

The values in the parentheses () are *p*-values; the asterisks ***, **, and * denote the level of significance of effect. If the p-value of Hausman specification test is above 5%, random effects (RE) model is employed; otherwise, fixed effects (FE) model is employed

effect on mortality rate. The testing results, based on the panel data of 20 provinces between 1995 and 2005, are as follows (see Table 3.3).

The primary focuses here are the regression results between 1998 and 2002. In model (1), the two dummy variables *Native* and *Tenure*5 are taken as proxy variables of collusion. After endogenous samples are eliminated, the attributes of the vice governor are exogenous, logically not subject to explanatory and control variables such as LnFixasset and Law_rate and this guarantees the unbiased estimation of key explanatory variables of collusion. The results based on random effects model show that *Native* and *Tenure5* significantly raise the mortality rate, which supports the assumption of collusion. Interestingly, after the variables of collusion are controlled, the coefficient symbols of the explanatory variables such as *LnFixasset, LnWage_per, LnRgdp_per, Law_rate*, and *Coal-rate* are consistent with what is theoretically predicted, but their effect on mortality rate is not significant.[9] The analysis above shows that collusion is the primary reason for high frequency of coalmine accidents, while capital input for safety, skill of coalmine workers, punitive enforcement, and economic development level are the secondary reasons.

In Model (2), the proxy variable *Age50* is employed and no essential difference can be found between the results in Models (1) and (2). The variable *Age50* significantly raise mortality rate and this shows that vice governors aged above 50 are more likely to collude with coalmine enterprises. This is because, compared with younger peers, they have fewer opportunities to be politically promoted and thus the cost of collusion is much lower.[10] The variable *Law_rate* significantly reduces mortality rate, which shows that raising judicial expenditure and strengthening punitive enforcement are conducive to reducing coalmine accidents.

In Model (3), three proxy variables *Native, Tenure*5, and *Age*50 are all taken into consideration and the results are almost the same as those in Model (1). It is noted that, if compared with Model (2), *Age*50 is not a significant factor for the following two reasons. First, the attributes of *Tenure*5 partially contains the information of *Age*50.[11] Secondly, the birthplace (*Native*) is comparatively more influential on mortality rate than age (*Age*50). The second reason is easily comprehensible in that China is a society bearing a tint of provincialism where people value the relationship between natives. In addition, after the variable *Age*50 is controlled, the coefficients of *Native* and *Tenure*5 are slightly altered and the coefficient of *Native* becomes bigger, which shows that the proxy variables in the explanation of government-firm collusion are robust.

Although Models (1), (2), and (3) reveal the significant effect of collusion on coalmine mortality rate, there is still a theoretical drawback: a variable *Z* that truly decides the mortality rate might be neglected and, moreover, *Z* might be in positive correlation with the two proxy variables *Native* and *Tenure*5 whose coefficients are biased. For this doubt to be erased, we draw two subsamples based on the 1995–1997 and 2003–2005 data. As is mentioned above, in the former phase, state-operated key coal enterprises were under the charge of the central administration and no collusion incentive existed between provincial levels and enterprises; in the latter phase, the State Administration of Coal Mine Safety was established and therefore collusion, to a certain degree, was contained. In the two subsamples, if it is theoretically supposed that collusion had nothing to do with coalmine mortality rate and that testing results also supported this assumption, then no variables would be neglected and the estimators between 1998 and 2002 would be unbiased.

In Model (4), the coefficients of the two proxy variables *Native* and *Tenure5* are not significant, and the coefficient of *LnFixasset* is obviously negative. This shows that the major reason for death rate is insufficient of capital input for safety before state-operated key coal enterprises were delegated to provincial governments. In Model (5), the coefficients of *Native* and *Tenure5* are still not significant, but the coefficient of *Coal_rate* is significantly negative. Although the coefficient symbols of the proxy variables in collusion are still consistent with what is theoretically predicted after the State Administration of Coal Mine Safety was founded, they are not significant and *Coal_rate* is the leading factor. This reveals that the SACMS plays a significant role in reducing the death rate of coalmine workers. The results of Models (4) and (5) show that no decisive or key variables are neglected in Models (1), (2), and (3) and that the estimator is unbiased.

3.5 Robustness Test

3.5.1 More Control Variables

For the effect of the tenure of a vice governor on coalmine death rate to be observed more minutely, a series of dummy variables are added to the 10-year tenure of vice governorship (supposing that a vice governor wins a second term of office consecutively) in Model (6) on the basis of 1998–2002 sample. Table 3.4 shows that the proxy variable Native is still positive, but the coefficients of *Tenure5* and other years of tenure are not significant. Considering altogether there are 71 observed values while there are only 15 explanatory variables, it is normal for the coefficients of different years of tenure to be not significant. When comparing the unbiased effects from the first year to the tenth of the tenure, it is easy to observe that the death rates in the fifth, sixth, and seventh years of the vice governor are usually higher than in other years. Based on repeated games and private contracts, it proves, from another perspective, a

Table 3.4 Contributors to death rates of state-owned key coal enterprises (2)

Attributive variable deathrate				
Phases	*1998–2002*			*1995–2005*
Model	*(6) RE*	*(7) FE*	*(8) OLS*	*(9) RE*
Native	4.920**	4.284*	4.760**	2.018
	(0.042)	(0.059)	(0.032)	(0.035)
Tenure5	8.201	5.906**	5.578*	2.890**
	(0.232)	(0.042)	(0.054)	(0.571)
Exper	1.013	0.709		
	(0.819)	(0.874)		
Native_g		1.864	0.398	
		(0.434)	(0.871)	
LnFixasset	−1.064	−0.882	−0.936	−0.619**
	(0.087)	(0.180)	(0.098)	(0.032)
LnWage_per	−5.931	−5.622	−4.305	−2.590
	(0.298)	(0.318)	(0.423)	(0.310)
LnRgdp_per	0.476	1.371	0.966	0.907
	(0.856)	(0.635)	(0.727)	(0.468)
Law_rate		−0.797	−2.098	−1.686
		(0.858)	(0.603)	(0.425)
Coal_rate		−3.256	−3.380	−5.407
		(0.634)	(0.602)	(0.066)
Y1998				3.883**
				(0.006)
Y2003				0.601
				(0.726)
Constant term	50.304	43.492	37.037	19.735
	(0.238)	(0.290)	(0.354)	(0.301)
R^2	0.243	0.231	0.136	0.190
Number of observed values	71	71	71	161

Notes The values in the parentheses () are a *p*-value; the asterisks ***, **, and * denote the significance levels of 1, 5, and 10% respectively. If the p-value of Hausman specification test is above 5%, random effects (RE) model is employed; otherwise, fixed effects (FE) model is employed

Chinese traditional belief of "acquaintances making lighter work". In view of the fact that a provincial decision in China is usually made by a governor in the form of centralization, a dummy variable *Native_g* (whether the governor is a native or not) is added in Model (7), and no essential difference can be found between the results of Models (7) and (1). This shows that the effect of the provincial government on coalmine safety is actually that of the vice governor on coalmine safety. Therefore, it is reasonable to choose *Native_g* as a proxy variable of collusion.

Model (8) reports the regression results based on the method of ordinary least squares (OLS). The coefficients of key explanatory variables *Native* and *Tenure5* are extremely significant. In Model (9), for the policies in different phases to be better controlled so as to solve possible omission of other variables, we define two dummy variables *Y1998* (denoting the Year of 1998) and *Y2003* and make a regression analysis of all samples. Besides that the coefficients of the two key explanatory variables are still significant, the coefficient of *LnFixasset* is significantly negative and the coefficient of the dummy variable *Y1998* is significantly positive. This shows that after 1998 some systematic factors and insufficient capital input for safety contribute to the increase of coalmine death rate.

In addition, between 1995 and 2005, 11 vice governors were abnormally removed from office or transferred to other lower positions (not transferred to positions with the same vice ministerial level). It is believed that their successors are endogenous. We re-add those formerly excluded endogenously observed value and find the regression results are still consistent with those from the major model.[12]

It is argued that increased coal price makes coalmines profiteering, and thus many coalminers' risks in reducing capital input for work safety and in expanding coalmines in frenzy contribute to higher mortality rates. However, when the national annual coal price index[13] is deliberately controlled, there is no substantial change for the major results. The likely reason is that coal price is positively correlated with coal output which is closely related with fixed asset investment.

3.5.2 *A More Rigorous Test*

Although the above quantitative analyses pass some robustness tests, there are still potential problems about the designing and testing of quantitative equations. These problems have been essentially solved by Jia and Nie (2017). They proved, by designing a mathematical model, their assertion that decentralization is the foremost important reason for local governments' collusion with state-run key coal enterprises. Moreover, they employed fixed effects model instead of random effects model. In addition, they not only explored the death rate changes of state-run key coal enterprises in different provinces in different provinces, autonomous regions, and municipalities but also compared administrative performances of vice governors of different provinces in different years. In the end, they added the degree of media disclosure of different provinces and other microdata concerning the relevant state-operated coal enterprises.

3.6 Summary

This chapter, based on the provincial-level dataset between 1995 and 2005, tests the effect of the collusion between local governments and state-run coal enterprises and of other factors such as safety input, coalmine workers' labor skill, and punitive enforcement on mortality rate. The empirical analysis shows that government-firm collusion is the culprit for coalmine death rate and that government–firm collusion will not have a significant effect on death rate any longer when the State Administration of Coal Mine Safety has come into its own as an independent supervision body free from provincial governments both economically and administratively.

The policy implications of this chapter are as follows: First, the central government should deter collusion, reduce the frequency of coalmine accidents and mortality rate, and make sure the State Administration of Coal Mine Safety retains its independence of interventions from provincial governments. Second, it is advisable that the officials responsible for coal mine safety in different levels of local governments (from provinces down to prefectures, cities, counties, and townships) should be non-natives and that their tenure should be strictly limited within 4 years, for instance. This conclusion has also certain significance on problems in public governance, such as environmental protection, quality supervision, and anticorruption.

Notes

1. See http://www.chinasafety.gov.cn/newpage/, the official website of the State Administration of Work Safety and BP Global Energy Statistics 2008 (Accessed June 18, 2011).
2. See http://www.chinasafety.gov.cn/newpage/, and the official website of the United States Department of Labor Mine Safety Health Administration (Accessed June 18, 2011).
3. Because pernicious accidents occurred thick and fast in these small coal pits, the State Council demanded that they should be closed by the end

of 2001. See Regulation No. 2001 [25] promulgated by the General Office of the State Council on June 30, 2001.

4. See "Delegating State-Owned Key Local Enterprises to Local Governments", *The People's Daily*, September 7, 1998.
5. At the provincial level, although major decisions are usually made by the provincial standing committees of both the CPC and the government, the vice governor in charge of a particular sector (such as agriculture, industry, education, health, etc.) has a weighty decision on these major issues. Since it is impossible to directly observe the relationship between the vice governor and coal enterprises, we can merely surmise about the probability of government-firm collusion through economics theory.
6. When observing the relationship between political promotion and economic performances, the descriptive statistics made by Li and Zhou (2005, p. 1753, Table 2) show that ministerial officials are likely to be promoted at 55. By analogy, the assumption that vice-ministerial officials are likely to be promoted at 50 is also qualified.
7. It is believed that high frequency of coalmine accidents may put major provincial leaders into jeopardy, but this does not mean collusion will never happen because coalmine accidents always happen with a certain degree of probability even if a coal pit is undermanaged and neglected. Furthermore, there is not always a one-to-one correspondence between the vice-governor's behaviors and coalmine accidents, which as a matter of fact multiplies loophole opportunities for local officials to participate in collusion.
8. According to *Coal Mine Safety Supervision Regulations of 2000* (partially revised in 2013), the highest financial penalty against a coalmine enterprise is CNY 150,000. According to the Criminal Law of the PRC, the severest criminal penalty against it is 7 year imprisonment.
9. When *Native* and *Tenure*5 are not controlled, *Law_rate* is significantly negative.
10. According to a report conducted by the Chinese Social Sciences Academy, 50% of local officials aged 55 to 60 have ever thought of abusing authority to skim off profits and take bribery before retirement and over half of them were indicted on corruption charges at the age of 59. See http://news.ifeng.com/mainland/special/fanfu/content-3/detail_2014_02/25/34138962_0.shtml (Accessed by translator on December 26, 2016).
11. Through regression analysis, *Tenure*5 and *Age*50 are in highly positive correlation.
12. After the 26 excluded observed values were re-added, the coefficient (p-value) of *Native* is 3.882 (0.044) and the *p*-value of *Tenure*5 is 3.684 (0.063), respectively significant above the level of 5 and 10%.

13. At present, there have been no regional coal price indices (price differences exist between different regions within the vast country) and power coal indices have not yet been fully compiled.
14. It is generally believed that, in the regression analysis of panel dataset, the results of fixed effects model are more reliable than those of random effects mode.

References

Cao, Yu, Zhou Zhou, and Di Wen. "Factors that Affect Coalmine Workers' Sense of Security in Hunan Province and Improvement Strategy". *Career Horizon*. 2007, 3(5): 58–61. 曹渝, 周舟, 文迪. 湖南煤矿工人心理安全感的影响因素及提升策略. 职业时空, 2007, 3(5): 58–61.

Faccio, Mara. "Politically Connected Firms". *American Economic Review*. 2006, 96 (1): 369–386.

Fisman, Raymond. "Estimating the Value of Political Connections". *American Economic Review*. 2001, 91: 1095–1102.

Heston, Alan, Robert Summers and Bettina Alten. *Penn World Table Version 6.2*, Center for International Comparisons of Production, Income and Prices at the University of Pennsylvania, September 2006.

Jia, Ruixue, and Huihua Nie. "Decentralization, Collusion and Coalmine Deaths in China". *Review of Economics and Statistics*. 2017, 90 (1): 105–118.

Laffont, Jean-Jacques. *Incentives and Political Economy*. Oxford: Oxford University Press. 2000.

Li, Hongbin and Li-An Zhou. "Political Turnover and Economic Performance: The Incentive Role of Personnel Control in China." *Journal of Public Economics*. 2005, 89: 1743–1762.

Nie, Huihua and Jinbo Li. Collusion and Economic Growth. *China Economic Quarterly*. 2006, 6 (1): 75–90. 聂辉华, 李金波. 政企合谋与经济发展. 经济学(季刊). 2006, 6(1): 75–90.

Nie, Huihua and Jinbo Li. "Collusion and Growth: A New Perspective on China Model". *Economic and Political Studies*. 2013, 1 (2): 18–39.

Tang, Lingxiao and Guo Xibao. "Causes and Countermeasures Against Frequent Occurrence of Coalmining Accidents in China Today: Based on the Standpoints of Safety Inputs". *China Industrial Economics*. 2006, 21 (12): 39–56. 汤凌霄, 郭熙保. 我国现阶段矿难频发成因及其对策: 基于安全投入的视角. 中国工业经济. 2006, 21 (12): 39–56.

Tirole, Jean. "Collusion and Theory of Organizations", in Jean-Jacques Laffont, eds. *Advances in Economic Theory: Proceedings of the Sixth World Congress of the Econometric Society*. Cambridge: Cambridge University Press. 1992.

Tirole, Jean. "Hierarchies and Bureaucracies: On the Role of Collusion in Organizations". *Journal of Law, Economics, and Organization*. 1986, 2 (2): 181–214.

Wang, Shaoguang. "Regulating Death at Coalmines: Changing Mode of Governance in China". *Journal of Contemporary China*. 2006, 15 (46): 1–30.

Zhong, Xiaohan. Coal Mine Security and Coal Production: A Tentative Study. Working paper of Qinghua University, School of Economics and Management. 2007. 钟笑寒. 煤矿安全与煤炭产量: 初步研究. 清华大学经济管理学院工作论文, 2007.

CHAPTER 4

Collusion and House Prices

This chapter is mainly concerned with measuring collusion with three measurement indicators: the proportion of land transaction fees (or land grant fee) to fiscal expenditure, state capital of housing industry, and indemnificatory (compensatory) housing investment. Section 4.1 gives a brief overview of exorbitant housing prices since the mid-1990s. Section 4.2 makes a literature review about the origins of exorbitant housing prices and provides a neopolitical economic perspective to explain this phenomenon. Section 4.3 introduces a relevant institutional background about regulating housing market. Section 4.4 expounds the major viewpoint of this chapter: government-firm collusion jacks up housing prices. Section 4.5 tests the relationship between collusion and housing prices by using the 2001–2010 data associated with commercial residential houses of all 31 provinces, municipalities, and autonomous regions in Mainland China. Section 4.6 is a summary.

4.1 An Overview of Exorbitant House Prices in China

Exorbitant house prices, in the wake of high tuition fee and staggering medical charge, is humorously dubbed as a third new big mountain[1] pressing on the contemporary Chinese people. This fully displays the formidable challenges daunting China in the provisions of public services. "Mortgage slave" has become an epithet of debt-ridden home buyers or proprietors. Suffice it to say that in China the issue of house prices has become the paramount concern of livelihood issues.

H. Nie, *Collusion, Local Governments and Development in China*,
DOI 10.1007/978-981-10-5059-6_4

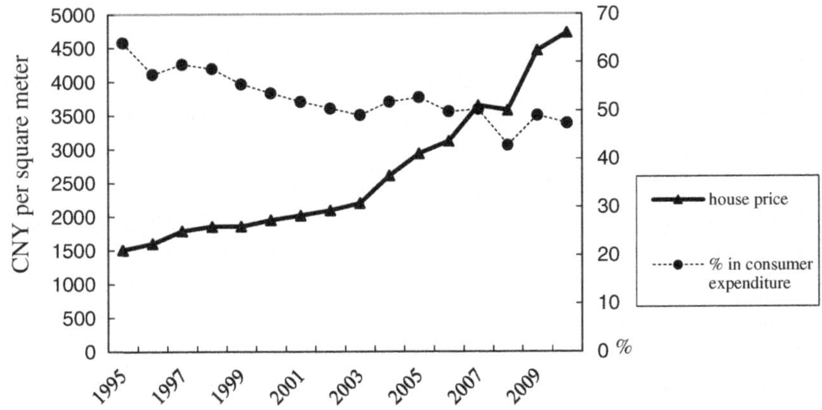

Fig. 4.1 Average house price and its percentage in consumer expenditure

Since the implementation of tax-sharing system reform in 1994, especially of the housing allocation reform in 1998, prices of commercial–residential buildings across the country have been on the increase. Figure 4.1 shows the average house price and its percentage in consumer expenditure. After a steady rise from CNY 1500 per square meter (about 10.76 square feet) of floor space in 1995 to CNY 2000 per square meter in 2003, house prices began to shoot up. By 2010, the national average house price was CNY 4725 per square meter. After computing the proportion of house prices to household consumption expenditure over the past 15 years, we find the figure was always high above 50%, and even drew near to the mark of 70%. To be more specific, most Chinese spent more than half of their income on housing. In addition, China's housing price-to-wage ratio was also much higher than that of global average.

Of the major cities in China, Hangzhou, Beijing, and Shanghai ranked top three in terms of house prices. According to *Ranking List of House Prices of Major Cities in China* (Table 4.1) issued in July 2010, the average price for a newly built yet unfurnished house in Hangzhou was CNY 25,840 per meter of floor space. On the heels of Hangzhou was Beijing with its house prices averaging CNY 22,310 per square meter. Shanghai was in the third place, averaging CNY 19,168 per square meter.[2] The house price of Hangzhou was eight times more than that of Karamay (CNY 3200) which ranks 100th place in the list. Consider, a house with 100 square meters of floor space would cost a

Table 4.1 Ranking list of house prices of major cities in China (July 2010)

Rank	*City*	*Unit price (CNY per m²)*	*Rank*	*City*	*Unit price (CNY per m²)*
1st	Hangzhou	25,840	26th	Quanzhou	7680
2nd	Beijing	22,310	27th	Jiaxing	7587
3rd	Shanghai	19,168	28th	Haikou	7288
4th	Wenzhou	18,854	29th	Dongguan	7023
5th	Sanya	18,319	30th	Chengdu	6630
6th	Shenzhen	16,978	31st	Hefei	6255
7th	Ningbo	13,438	32nd	Erdos	6220
8th	Guangzhou	12,560	33rd	Wuhan	6196
9th	Nanjing	12,016	34th	Nanning	6137
10th	Zhoushan	10,500	35th	Tangshan	6098
11th	Shaoxing	10,105	36th	Kunming	6006
12th	Zhuhai	9747	37th	Jinhua	6000
13th	Dalian	9678	38th	Zhongshan	5998
14th	Xiamen	9660	39th	Yantai	5835
15th	Suzhou	9103	40th	Chongqing	5720
16th	Taizhou	8967	41st	Rizhao	5700
17th	Qingdao	8962	42nd	Zhengzhou	5689
18th	Tianjin	8958	43rd	Huizhou	5649
19th	Nantong	8950	44th	Yichang	5637
20th	Fuzhou	8666	45th	Taiyuan	5635
21st	Huzhou	8220	46th	Yangzhou	5630
22nd	Harbin	7939	47th	Nanchang	5573
23rd	Wuxi	7843	48th	Changchun	5445
24th	Foshan	7822	49th	Lanzhou	5440
25th	Jinan	7760	50th	Xi'an	5398

Source Livable city research team, July 2010
http://www.360doc.com/content/10/0716/23/554737_39538326.shtml. (Accessed March 19, 2017 by the translators)

typical Beijinger with an average annual income of CNY 50,145 44 years in wages even if he paid in cash to purchase it without sparing a single coin on food and drink.[3] Suppose, he paid a downpayment of 30% and applied for a 30-year-long (the longest repayable period) commercial loan for the remaining 70% with an interest rate of 6.55%, the total amount of repayment would amount to CNY 3,572,069.35 or monthly repayment CNY 9,922.41, twice as much as his monthly wage.[4] It is obvious that an individual with an average annual income could hardly afford a house in Beijing.

Table 4.2 The Central government's polices regulating houseprices

Date	*Relevant policies or regulations or other measures*
March 26, 2005	The State Council issued *Circular of the General Office of the State Council on Effectively Stabilizing HousePrices*, cracking down speculation in real estate
April 27, 2005	The State Council added another eight measures to curb house prices, with the emphasis on tightening control over land, building affordable houses for low-income families and guaranteeing the supply of low-rent houses
May 24, 2006	The State Council promulgated six measures about readjusting the supply structure of houses, focusing on development of middle- and low-price houses, small- and medium-dwelling unit type houses, etc.
2007	The central bank (People's Bank of China, PBC) consecutively raised interest rates five times and reserve requirement ration (RRR) 10 times and constrained capital over-liquidity and overheating of house market, supporting low-rent houses and low-income families
April 17, 2010	The State Council promulgated ten new measures to control irrational demand for houses, increase responsive supply chains, speed up constructions of affordable housing projects and strengthen market supervision
September 29, 2010	The State Council promulgated measures to suspend house loan for a family's third house. It demanded that the down payment of a family's first house should be no less than 30% of the purchase price and that cities with fast-increasing house prices should limit the number of houses a family could afford
January 26, 2011	The State Council demanded that the down payment of a family's second house should be no less than 60% of the purchase price, and interest rate of house loan should be at least 1.1 times more than benchmark interest rate

Sources Rearranged according to www.gov.cn, website of the Chinese Central government and www.pbc.gov.cn, website of People's Bank of China. (Accessed August 5–6, 2011)

To scale back accelerating house prices to an affordable level for ordinary residents, the Chinese central government successively formulated a series of policies about regulating house prices (see Table 4.2). On March 26, 2005, the State Council issued *Circular of the General Office of the State Council on Effectively Stabilizing House Prices,* which included eight measures. On April 27, 2005, the State Council added another eight measures to curb house prices. On January 26, 2011, the

Executive Meeting of the State Council promulgated for the third time eight measures to squeeze out speculative binge and dampen the soaring house prices, demanding local governments at different levels throughout the country to issue "House Purchase Restriction," an unprecedentedly tough order to limit the number of houses that one family can purchase. More than 70 large and medium cities followed suit, but house prices were still moderately rising (Nie and Li 2011).

4.2 Three Different Views About Exorbitant HousePrices

There is a thought-provoking question: why have house prices remained so high without any sign of decline in such a centralized authoritarian state, where the central government has been in the grip of housing reform fever attempting to bring it down in recent years? Paradoxically, the central government can effectively control prices of other commodities such as oil and vegetables. A reasonable interpretation of this phenomenon necessitates the following two facts: on the one hand, the central government holds the supreme political authority to appoint and remove officials at the provincial level; on the other hand, the governments at different levels intervene in the real estate market; so, house prices are affected by governmental decisions. Much of the existing literature about housing market ascribe high house prices to macro-policy and market mechanism (supply and demand), but this is far from being enough to give a satisfactory explanation.

There were three viewpoints as to China's current exorbitant house prices. The first viewpoint held that supply and demand determined house prices. For instance, Liang and Gao (2007) observed that real GDP per capita and price expectation had a great impact on house prices within a short term. Kuang (2010) analyzed the data collected from 35 large- and medium-sized cities and concluded that residents' expectation of prices and their speculative binge could effectively interpret the fluctuation of urban house prices. Zou and Liu (2010) analyzed the urbanization data from 1987 to 2006, observed that the fluctuation of house prices could be explained by GDP per capita and the invariables of construction cost, and their observation was shared by Chen and Fu (2011). Lu (2011) noted that it was governmental monopolistic supply of land that jacked up house prices. From the perspective of neoclassical economics, price is determined by the two forces of supply and demand. However, the question is: Why does the local government that controls the supply of land not reduce the house prices by increasing the supply of land (e.g., allowing collective land to enter into the secondary market) on the one hand, and by reducing the demand for commercial houses (e.g. increasing residential houses for medium-and-low wage earners and low-rent houses) on the other? It is evident that the question cannot be satisfactorily solved without a good understanding of the conflict of interests between the central government and the local governments.

The second view believed that it was the fiscal or monetary policies that pushed up house prices. Han and Kuang (2012) held that local governments tried every conceivable means to allocate land resources to industrial sectors with more sources of tax so that it could gain more fiscal revenue and furthermore they tested their observations by using 1999–2006 data about prefecture-level cities. Lü and Nie (2014) pointed out that it was very natural for local governments to stimulate housing industry on which both value-added tax and sales taxes (double taxation) were levied so as to ease the fiscal pressure after 1994 tax-sharing system reform. Li and Li (2012) testified to the positive correlation between fiscal decentralization and high house prices. Wu (2010) held that the reason for quick increase of urban house prices was the liquidity of housing market, especially the flood of a large amount of credit fund into the housing market due to the CNY4 trillion ($586 billion USD) stimulus package in November 2008. Xu (2011) argued that it was the lowered real interest rate due to inflation that substantially increased the demand for house purchase for investment purpose and consequently jacked up house price. Nevertheless, these ideas mentioned above could not answer this question: if the fiscal or monetary policies should be blamed for high house price, why did the central government (here represented by the Central Bank) responsible for fiscal or monetary policies not regulate the policies accordingly?

The third view transcended the framework of neoclassical economics, arguing that the collusion between the local government and real estate enterprises was the underlying reason for high house prices in China. For instance, Su and Tao (2010) argued that the local government, commercial banks, and real estate enterprises formed a stable coalition of interests for the purpose of gaining more land transaction fees and therefore this alliance lacked sufficient incentives to execute the central government's orders and policies concerning house price control. Nie and Li (2011) noted that the 1994 tax-sharing system weakened the local government's fiscal strength and thus it colluded with real estate enterprises to push up house prices so as to gain more fiscal revenue from land. Though the third view noted the conflict of interest between the central and local governments, no literature provided a solid microscopic or a systematical empirical test.

This chapter provides a neopolitical economic perspective to explain high house prices. It is believed that in the game where the central government, the local government, and real estate enterprise are participants, the central government is confronted with a tradeoff between economic growth and social stability. The central government is not informed of the real cost that the local government develops its local economy, but it should bear the ultimate cost of safeguarding social stability. In order to be economically rewarded and politically promoted, the local government and real estate enterprises take advantage of information asymmetry to collude in pushing up house prices and GDP (measurement of political achievement), and hence profiteering from it by means of monopolizing land supply, providing capital, and increasing the demand for houses. Since the central government is heavily dependent on economic growth, on the one hand, but is afraid that high house prices will cause social instability, it periodically intervenes in the house market. By analyzing the data of commercial residential houses in the 31 provinces, autonomous regions, and municipalities from 2001 to 2010 and using the three indices—the percentage of land transaction fees in fiscal expenditure, state capital in housing industry, and indemnificatory housing investment—to measure government–firm collusion, we conclude that collusion noticeably increases the price of commercial residential houses. More importantly, government–firm collusion can explain almost all changes of house prices.

This chapter introduces the behaviors of central and local governments and provides a microscopic basis on which to understand high house prices rather than merely depending on the stereoscopic perspective of supply and demand. Though mass media and economics literature mentioned the impact of collusion on house prices, it is the first time for us to test it with the data collected from the provincial level.

4.3 The Institutional Background of Housing Market Control

4.3.1 Fiscal Decentralization Between the Central Government and Local Governments

The fiscal relationship between the central and local governments has undergone the following stages. From 1949 to 1979, China implemented centrally-unified collection and allocation of taxes, thus discouraging local governments from fostering tax bases and levying taxes because they had no residual claim rights. From 1980 to 1983, China put into practice fiscal responsibility system, under which the central government was not well-advised of accurate information of tax bases owing to information asymmetry and hence the local government had the incentive to underreport tax bases so as to retain a large amount of fiscal revenue by decreasing the actual percentage of taxes which it was required to turn over to the central government. On the other hand, if the central government discovered that local governments underreported tax bases, it would increase the taxation quota and in turn ratchet effect followed (Lü and Nie 2014). Therefore, the fiscal responsibility system inevitably had a perverse effect on national macroscopic readjustment and control because it resulted in a double drop of the proportion of national fiscal revenue to the GDP aggregate and the proportion of the central government's fiscal revenue to total national fiscal revenue. It was this double drop that led to the reform of tax-sharing system in 1994. Under the tax-sharing system, the proportion of central tax revenue increased, whereas the proportion of local tax revenue decreased accordingly. However, since the local governments' duty of undertaking fiscal expenditure did not lessen accordingly, such an unequal fiscal relationship of "centralizing fiscal authority but decentralizing governance affairs" existed between the central government and local governments. Figure 4.2 shows that, prior to the 1994 tax-sharing reform, the local governments' proportions of fiscal revenue and expenditure to the national total remained balanced; but after 1994, the proportion of fiscal revenue decreased sharply from 75% to about 45%, and since then fluctuated between 45 and 50% while the proportion of fiscal expenditure steadily increased from 65% to more than 80%.

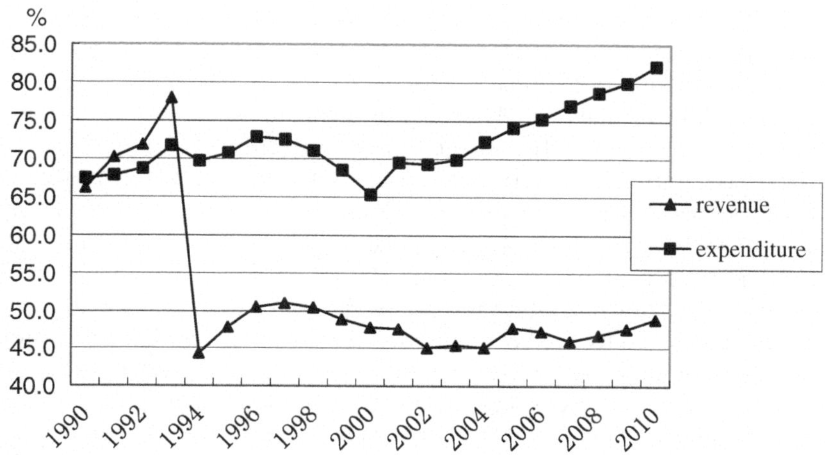

Fig. 4.2 Proportions of local governments' fiscal revenue and expenditure

4.3.2 Housing and Land Reform

After the founding of the People's Republic of China in 1949, the central government established work unit (*danwei* in Chinese *pinyin*) system in cities and townships. Work unit, a place of an individual's employment and a principle of implementing CPC's policy, functioned as the first step of multitiered hierarchy linking each individual with Central Party. Before 1998, city and township dwellers enjoyed social welfare housing and each of them could be distributed a welfare house or apartment room when he or she got employed by a work unit. Since the issue of the *Circular of the State Council on Further Deepening Housing System Reform in Cities and Townships and Quickening Housing Construction* on July 3, 1998, a fundamental change—from welfare-orientation to monetization—has taken place in housing allocation reform. From then on, city dwellers became the main purchasers of home buildings. With accelerated urbanization and increased income, the city dwellers' demand for commercial residential buildings was greatly released (Zou and Niu 2010). Furthermore, the housing reform promoted housing supply, financing, and intermediary services so that the State Council specified in the *Notice on Promoting the Stable and Sound Development of the Real Estate Market* in 2003 that housing industry was rated as a pillar industry in the construction of national economy.

Despite the substantial release of the demand for housing, housing supply had been rigorously regulated by the government.

To begin with, the government controlled land supply. For instance, Article 43 of Chap. V "Land for Construction Purposes" in *Land Administration Law of the PRC* stipulates: Any unit or individual that needs land for construction purposes should apply for the use of land owned by the State according to law, except land owned by peasant collectives used by collective economic organizations for building township enterprises or building houses for villagers or land owned by peasant collectives approved according to law for use in building public facilities or public welfare facilities of townships. Article 63 stipulates: The land use right of peasant collectives shall not be leased, transferred, or rented for nonagricultural construction, except in the case of legal transfer of land that conforms to the general plan for the utilization of land and legally obtained by enterprises due to bankruptcy or acquisition. In other words, any land, be it state-owned or collective, shall be requisitioned by the government before entering into the secondary land market and finally being used for commercial purpose. In many cases, all old buildings on a piece of land to be commercially developed will be demolished to make room for new construction. House demolition tends to trigger off a series of social contradictions, such as the conflict of interests between families to be relocated and government departments which, according to law, shall be responsible for house demolition and relocation and for the distribution of compensation fees that have not totally been marketized. In secondary markets, the land held by the government is not supplied gratis but auctioned as required by marketization. In April 2001, the *Notice of the State Council on Strengthening the Administration of State-owned Land Assets* required that land use right bidding and auctioning should be vigorously promoted all the country, signifying a shift from nonmarket allocation to market allocation in the development of profit-oriented land. In the same year, the Ministry of Land and Resources and the Ministry of supervision jointly issued the *Notice on Supervision of Continuing Land Use Right Sales Through Bidding, Auction, and Listing* (No. [2004] 71), demanding that contracted yet unfinished land bidding, auction, and listing should be completed before the deadline (otherwise known as August 31 Deadline).

From then on, governments at different levels across the country raced to implement this practice. In August 2006, the *Circular of the State Council on Intensifying Land Control* specified that land for industrial purposes should also be leased through bidding, auction, and listing.

Second, housing developers virtually monopolized the supplies of housing. In the practice of land bidding, auction, and listing, it is those housing developers with a strong financial background can access land tenure, the foremost important prerequisite for such subsequent procedures as housing construction, engineering supervision, and housing presale. On the other hand, collective property right houses currently held by rural and township residents (commonly referred to as houses with limited property right), unlike commercial residential houses, are forbidden to enter into the distribution channel. This category of house exists in a grey area between legal and illegal.

4.4 High House Prices Driven up by Collusion

Supposing there is a game where the central government (representing the principal), a local government (representing the supervisor), and a real estate enterprise (representing the agent) are participants (Nie and Li 2006). The aim of the central government is to boost economic growth and maintain social stability. The local government, entrusted by the central government, stimulates regional economic prosperity by means of regulating the development of housing industry. With economic growth, the local government is rewarded with bumper fiscal revenue and political promotion. The local government has the following two features. First, it has well in hand the private information about economic growth. Second, its short tenure (according to the Constitution, there shall be a quinquennial reshuffle in the local government) implies that it pays little heed to the cost of stabilizing social order that might be concomitant with economic growth. Intrinsically, the said real estate enterprise is a production unit regulated by the local government, so the information is symmetrical between them. Given a target of economic growth, the local government has two opposite strategies in dealing with the real estate industry: encouragement versus restraint. Encouragement inflates house price, jacks up land value, increases GDP aggregate, blows economic bubble, and widens income gap while restraint aggravates corruption and eventually roils social stability.[5] However, restraint maintains house and land prices at a moderate level, balances resident income level, and sustains economic growth at a steady rate.[6]

Chronologically, the sequence of the game is as follows. First, the central government makes a grand contract and establishes policies about economic development and maintenance of social stability, including the regulations concerning reward and punishment mechanisms targeted at local officials and the real estate enterprise. Second, the local government chooses a strategy for economic development; under the condition of information asymmetry, it colludes with real estate enterprises and chooses the aggressive strategy. Third, given polices promulgated by both the central and local governments, the real estate enterprise chooses the action of profit maximization.

Then, in a game equilibrium, what are the optimal strategies chosen by the central government, the local government, and the real estate enterprise respectively? Based on the mathematical model proposed by Nie and Li (2006, 2013), two corollaries can be drawn as follows:

Corollary 4.1 *If the potential profits and gains arising from collusion exceed possible losses, or the likely punitive measures are far less than expected, the local government colludes with the real estate enterprise, encouraging the development of real estate industry, pushing up house prices, land value, and GDP aggregate.*

Corollary 4.1 shows that collusion may exist as an equilibrium. The central government makes a judgment as to a local government official's governance and competence simply through his or her easily observable results—economic output and social stability—because it is not familiar with other seemingly negligible achievements such as the improvement of medical care and education that he or she has made. In the case of social stability, the central government has a tendency to review local officials and decides their political mobility—promotion, demotion, or removal—and fiscal reward through GDP aggregate and fiscal revenue. Under this incentive mechanism, given that the central government metes out merely a slight punishment to collusion between the local government and the real estate enterprise or that the probability for the central government to perceive the collusion is very low, the local government is very likely to choose the aggressive strategy so as to get a higher probability of political promotion (Li and Zhou 2005) and a higher proportion of fiscal revenue (Qian and Xu 1993; Jin et al. 2005).

It is noteworthy that the term *collusion* in the game theory refers to an improper secret agreement between the supervisor and the agent to form a side contract at the expense of the principal's interests. As a matter of fact, the central government, although eager to achieve its target, is under no circumstances willing to push up house price, land value, and GDP aggregate without taking social stability into consideration, so much so that it issued in quick succession notes, circulars, and many other measures (see Table 4.2) to curb house prices since 2005. Moreover, in late-July 2012, the central government sent 8 inspection teams to 16 provinces, municipalities, and autonomous regions to investigate the execution of policies concerning real estate control issued by the State Council, especially of *House Purchase Restriction*.[7] However, the local government is only a short-term supervisor[8] who cares much more about short-term economic growth targets and political mobility than about the medium- and long-term adverse effects of real estate bubble on social stability and adjustment of industrial structure. As the subject of being regulated for profit maximization, the real estate enterprise will collude with the local government only when the latter chooses the aggressive strategy. Otherwise, it is reluctant to do so because it cannot get any profits in the case of conservative strategy.

Against the background of fiscal decentralization, the local government is motivated to collude with the real estate enterprise for the following reasons.

First, as mentioned above, since the tax-sharing system reform in 1994, the local government has been delegated more affairs to deal with but given less authority to decide how much percentage of tax revenue can be retained and this reform inevitably resulted in the imbalance of financial revenue and expenditure. Under the tax-sharing system, the main portions of the two biggest tax categories—value-added tax and income tax—should be turned over to the central government. At present, land transaction fees are the main source of non-tax revenue for the local government. Table 4.3 shows that land transaction fees accounted for only 16.6% of the local governments' fiscal revenue quickly in 2001, but quickly rose to 76.6% in 2010. In other words, nearly 80% of the local governments' fiscal revenue was from land transactions. So to speak, "land finance" has sarcastically become the epithet of local fiscal revenue. From the perspective of regional distribution, the data of the year 2009 released by the Ministry of Land and Resources showed that in all the provinces, municipalities, and autonomous regions of mainland China, the average percentage of land transaction fees in fiscal revenue was nearly 50% but the highest arrived at 96%. A further calculation shows that the percentage of the land transaction fees gained

Table 4.3 Land transaction fees and local fiscal revenue

Year	*2001*	*2002*	*2003*	*2004*	*2005*	*2006*	*2007*	*2008*	*2009*	*2010*
LTF (billion CNY)	129.6	241.7	542.1	614.2	588.4	766.7	1300	960	1600	2700
LFR (billion CNY)	780.3	851.5	985	1189.3	1510.1	1830.4	2357.3	2865	3258.1	3538.3
Percentage of LTF in LFR	16.6	28.4	55	53.9	39	41.9	55.2	33.5	49.1	76.6

Source Zhang Yanjun et al. (2012)
Note 1 billion = 1000,000,000

from bidding, auction, and listing accounted for 30% of the fiscal revenue and the highest was 73%. (Nie and Li 2011) The underlying reason for the local government to get a large portion of land transaction fees is that the local government is the monopolistic land suppliers and lessors. Under the guise of protecting public interests, the local government encloses any piece of land of whatever commercial value by means of non-marketization such as house demolition and reallocation of state-owned land and then makes a huge profit by cashing out the enclosed land through marketization such as bidding, auction, and listing.[9]

Second, under political centralization and fiscal decentralization with Chinese characteristics, local officials are assessed and appointed by a higher level of state administrative organ instead of being voted by electorates. The main criterion of assessment is GDP aggregate or tax revenue (Bo 1996). Real estate industry benefits the local government with double taxation of value-added tax and business tax. In addition, it also booms other relevant industries such as iron and steel, cement, construction, and banking. Statistics of the China State Administration of Taxation show that taxes concerning real estate industry during the National Tenth Five-Year Plan (2001–2005) increased by 41.4% annually on the average. In 2008, transaction tax of real estate industry accounted for 22.5% of the total transaction tax of all industries (Lü and Nie 2012). To conclude, real estate industry is one of the chief industries which local government officials depend on to attain achievements in their political career.

The collusion between the local government and the real estate enterprise may be contrived in a variety of forms. First, the local government passively and superficially executes policies of readjusting and controlling real estate markets, but underhandedly issues measures to stimulate house purchase. For instance, during the 2 months after the issue of *House Purchase Restriction* in January 2011, only 40 of all the more than 600 cities published the target of readjustment and control, among which only Beijing announced its target of "steady and slight decline" while others announced their target of "steady rise".[10] Moreover, some local officials overtly declared that they would "prioritize the construction of small and medium cities rather than readjust and control real estate markets".[11] Second, the local government injects state capital into real estate enterprises, instigating developers to run up the land price when house prices fall and creating a false housing prosperity. Third, the local government promulgates relevant policies, pegging house purchase with the right to enjoy education, medical care, social security benefits, and other public services, and thus pushing up the demand for housing. Fourth, the local government ignores the construction of indemnificatory housing such as residential houses for low and medium wage earners and low-rent houses. Given a fixed amount of residential housing demand, the more indemnificatory house purchasers, and the fewer the residential house purchasers, the lower house prices. As a rule of thumb, the local government should supply more indemnificatory houses to control house prices as guided by the central government, but the reality is far from satisfactory. For instance, the central government planned in 2009 to allocate CNY 167,000 million to the construction of indemnificatory houses, but the local governments at different levels ensured merely CNY 39, 490 million, or 23.6% of the planned funds. Fifth, the local government lobbies some decision departments to relax stringent rules for real estate enterprises. For instance, during an investigative meeting of the Ministry of Housing and Urban-Rural Development in October 2009, representatives from the provinces of Jiangsu and Zhejiang expressed their hope to continue the lax housing measures (such as preferential tax policies) of 2009 in the coming year.[12] Sixth, the local government illegally zones arable land or public land for commercial purposes, or resells the land to real estate enterprises at a high price, or helps real estate enterprises evade taxes. Accordingly, estate enterprises will return a favor in exchange, collaborate with local officials to falsify their political performances, and even offer a bribe to them. To sum up, in the collusion between the local government and the enterprise, the two parties are bound for transfer of benefits. Without either party, the other could not push up house prices and profiteer from it. For instance, in a scandal disclosed by some muckrakers, Song Weiping, a real estate tycoon and billionaire and co-founder and president of Greentown China Holdings Limited, headquartered in Zhejiang province, greased the palm of Zhang Jiameng, Party Secretary of Zhoushan City (under the jurisdiction of Zhejiang province) and helped boost the local real estate industry; and in turn, the Party Secretary gave Greentown too much privilege in approving land use formalities and tax collection.[13]

In the collusion alliance formed by the local government and the real estate enterprise, there is another participant—the bank. A local commercial bank under the jurisdiction of the local government is usually regarded as an important entity inextricably linked with the local government. A branch of a vertically managed state-owned commercial bank (such as China Construction Bank and other Big Four Banks) is intrinsically the same as a real estate enterprise in that it is also in the pursuit of profit maximization. Theoretically, an absence of analyzing the bank as an agent will not fundamentally change the equilibrium result.

Corollary 4.2 *If the cost that the central government pays to stabilize society due to the collusion between the local government and the real estate enterprise is big enough, or the central government attaches more to social stability than to economic growth, the central government will deter collusion and enforce stringent regulations upon the real estate market. Owing to information time-lag, the central government observes ex post facto the cost arising from collusion. This signifies that the central government will intervene periodically in the real estate market.*

Corollary 4.2 analyzes the central government's decision. It is evident that the prerequisites for collusion between the local government and the real estate enterprise are: (1) the central government's time-lagged collusion-proof actions which lead to ex post facto intervention; (2) the central government's connivance at collusion.

In the first case, the central government colludes with the real estate enterprise to push up land and house prices and boost GDP growth, but concomitant with real estate bubble are the widening gap of income distribution and low-wage earners' resentment. It is the easily observable GDP growth data that catch the central government's attention first, but it usually takes a much longer time to recognize lurking aftermaths of economic bubbles. Therefore, the central government is not likely to punish the local government and the real estate enterprise. Instead, it may even award them by politically promoting them and fiscally giving them a larger share of taxes. This is the fundamental reason for collusion. When the central government has observed the negative effects of real estate rigging, it will stipulate readjustment and control regulations (such as the issue of *Circular of the General Office of the State Council on Effectively Stabilizing House prices* in March 2005) and incorporate the result of readjustment and control into assessment criteria of governmental officials. This explains why more stringent readjustment and control measures are hot on the heels of a sharp increase of house price, i.e., periodical intervention in real estate market.

In the second case, when economy is in dumps, the central government will stimulate economic growth instead of maintaining low house prices and even encourage the development of real estate industry. For example, when the global economy was in depression after the subprime mortgage crisis in 2007, the central government did not issue any policy concerning readjust and control of real estate market between 2008 and 2009 (see also Table 4.2).

In the following section, an econometric equation will be devised to test Corollary 4.2.

4.5 Results of Econometric Tests

4.5.1 Model Specification

Our aim is to test the effect of the collusion between local government and the real estate enterprise on house prices. In the econometric equation, the selling price of residential houses is treated as an explained variable. In terms of explanatory variables, it is empirically difficult to identify collusion in that it is a secret behavior of hidden information or act. Enlightened by Nie and Jiang (2011), we adopt an estimation strategy: when we have observed the occurrence of an accident, we believe that in real life the probability of collusion is actually even higher than what is observed, so these quantitative and qualitative accidents can serve as the proxy indicators of collusion. According to Corollary 4.1, the local government depends on land for fiscal revenue, so the variable *LandF* (land fiscal revenue) is picked to measure the contribution ratio of land transaction fees to fiscal expenditure. A higher ratio shows that the local government depends more heavily on land transaction fees or on the boom of real estate industry. This signifies that the local government has more incentives to collude with the real estate enterprise.

The benefit that the real estate enterprise can get from collusion is the generous profit yielding from exorbitant house prices; however, as to the profits that real estate enterprises get, we can by no means get the summed data at the provincial level. A real estate enterprise usually has a great demand for development funds so that it usually recoups funds by selling commercial residential houses in advance of construction. Therefore, we use *StateK* (denoting state capital) to measure the level of collusion between the local government and the real estate enterprise. This variable shows the funding amount the local government grants the real estate enterprise and the degree of importance which the local government attaches to the real estate industry. A higher sum of state capital signifies that local governments are more supportive of real estate industry and that the probability of collusion will be higher.

As stated above, the local government's attitude toward the construction of indemnificatory housing also indicates the level of collusion. Unlike commercial residential housing, indemnificatory housing refers to substandard or standard limited low-rent housing, economical housing, and public rental housing specially built for those with a medium and low household income. The land use rights for indemnificatory housing are remised by agreement through the local government and the transaction fee is relatively low. In view of this, the local government neglects the supply of indemnificatory housing, intentionally compelling an increasing number of consumers to purchase more costly commercial residential housing and supporting the real estate enterprise in a disguised form. Here, *EconH* (denoting economical housing investment)[14] can serve as a third proxy indicator. The less the variable, the less positive the local government's attitude toward the construction of indemnificatory housing, the more probable it colludes with the real estate enterprise to push up house prices.

Besides collusion, other factors, which are set as control variables for our purpose, affect house prices. First, urbanization in current China releases an increasing demand for housing (Xu 2012), so *Urban* (denoting urbanization) is chosen as a control variable. Second, the age structure affects the demand for housing (Mankiw and Weil 1989). It is generally believed that the working-age population is directly proportional to the demand for housing, so *Popu* (denoting dependency ratio of population) is also chosen as a control variable. Third, the supply of commercial residential housing influences the trend of house prices, so attention should also be paid to the control variable *Complet* (denoting completed floor area of commercial residential housing). Fourth, land use right sales through bidding, auction, and listing mentioned previously also push up land and house prices, so the control variable *Cost* (denoting cost of commercial residential housing) is used to refer to supply cost. Fifth, since a commercial residential house costs an average family a great percentage of income, the majority of families often purchase one by mortgage payment. This implies that banking interests influences both the behavior of house purchasers and the fund supply of real estate enterprises. (Carstensen 2009) Hence, we choose the variable *Rate* (denoting mortgage rates of 5 years and above) to control this factor. Eventually, according to Corollary 4.2, the central government's to regulate and control the real estate market depends on the global and Chinese domestic economic environment. To control the effect of this factor on house prices, we add the dummy variable γ (here denoting time fixed effect). Meanwhile, to control regional differences, we think it is necessary to add δ (here denoting place fixed effect).

To sum up, we establish the following econometric model of panel data:

$$\begin{aligned} LnHouseP_{i,t} = \alpha &+ \beta_1 Land_{i,t-1} + \beta_2 LnStateK_{i,t-1} \\ &+ \beta_3 LnEconH_{i,t-1} + \beta_4 Urban_{i,t-1} \\ &+ \beta_5 Popu_{i,t-1} + \beta_6 LnComplet_{i,t-1} + \beta_7 LnCost_{i,t-1} \\ &+ \beta_8 Rate_{i,t-1} + \delta_i + \gamma_t + \varepsilon_{i,t} \end{aligned}$$

To solve endogenous problems, we choose all explanatory variables lagged for one period as proxy variables. Meanwhile, to smooth data, we take the natural logarithms of these variables *HouseP*, *StateK*, *EconH*, *Complet*, and *Cost*. In the above equation, γ denotes time fixed effect, δ denotes place fixed effect and ε denotes errors.

4.5.2 Data Description

China began its housing reform in 1998. In view of availability, the sample time of the data is set from 2001 to 2010. The sample scope covers 31 provincial-level administrative units in mainland China. *LandF* (denoting land transaction fees) data are from *China Land and Resources Statistical Yearbook*, and *Rate* (denoting interest rate) data from "Loan Rates with a Term of Five Years and Above" issued by the website of the People's Bank of China. In the case of adjustments of rates in a certain year, we computed the overall interest rate level of that year in terms of the weighted average rate based on the numbers of the months when the old and new loan rates were respectively adopted. The data about *HousP*, *Complet,* and others are from *China Statistical Yearbook* and the *China Social and Economic Development Database* of China National Knowledge Infrastructure (CNKI). We have obtained 310 observations all told. Table 4.4 shows a descriptive statistics of major variables.

4.5.3 An Analysis of Regression Results of Housing Variables

In order to explore the effect of collusion on house prices, we add explanatory variables step by step and computed major regression results (see Table 4.5).

Table 4.4 A descriptive statistics of major variables of housing

Variables	*Unit*	*Mean*	*Standard deviation*	*Minimum*	*Maximum*	*Observations*
HousP	CNY/m^2	2828.65	2125.71	864	17,151	310
LandF	Ratio	0.19	0.194	0.001	1.31	310
StateK	Million CNY	4677.695	9795.681	2.73	90,382.31	275
EconH	Million CNY	2359.278	1962.162	10.21	11,189.27	301
Urban	Ratio	0.409	0.153	0.005	0.87	300
Popu	Percent	42.228	7.861	26.99	63.46	310
Complet	Thousand CNY	12,746.1	10,962.7	66.6	67,535.7	310
Cost	CNY/m^2	1347.79	522.61	644	3353	310
Rate	Percent	6.33	0.61	5.76	7.63	310

Table 4.5 Main regression results of housing variables

Explanatory Variables	*Dependent variable: Selling price of commercial residential housing*				
	OLS (1)	*FE (2)*	*OLS (3)*	*OLS (4)*	*FE (5)*
LandF	0.315***	0.291***	0.434***	0.410***	0.211*
	(0.122)	(0.119)	(0.135)	(0.101)	(0.119)
StateK	0.241***	0.049***	0.243***	0.110***	0.049***
	(0.017)	(0.013)	(0.019)	(0.016)	(0.014)
EconH			−0.068***	−0.034*	0.013
			(0.023)	(0.020)	(0.022)
Urban				0.784***	0.345***
				(0.132)	(0.101)
Popu				0.0003	0.004*
				(0.002)	(0.002)
Complete				−0.033	−0.043
				(0.023)	(0.052)
Cost				0.700***	0.165**
				(0.056)	(0.075)
Rate				0.081**	0.452***
				(0.038)	(0.065)
Region		Y			Y
Year		Y		Y	Y
R^2	0.639	0.967	0.623	0.894	0.968
Observations	275	275	267	259	259

Notes Robustness standard error is employed for all regression models. All values in the parentheses denote a standard error. The asterisks ***, **, and * denote the significance levels of 1, 5, and 10% respectively

Let us first examine the effect of *LandF* and *StateK* —two positive directional indicators that test the level of collusion—on house prices. Regression model (1) shows that *LandF* and *StateK* at the significant level of 1% significantly increase house prices if not other variables are not controlled. This is compatible with our theoretical prediction. Model (2) shows that the coefficients of the two variables *LandF* and *StateK* are still significant even if both *Region* (place fixed effect) and *Year* (time fixed effect) are controlled. R^2 is as high as 0.967, which indicates that collusion effectively explains skyrocketing house prices. In Model (3) where the third proxy indicator *EconH* is added, *EconH* significantly reduces house prices and the significance level of *LandF* and have only an almost imperceptible change. The coefficients of the three variables *LandF*, *StateK*, and *EconH* also correspond to our theoretical prediction.

In Models (4) and (5), we put all explanatory variables and control variables into regression equations and then we regress these variables by means of ordinary least square (OLS) and fixed effect (FE), respectively. In the two models, both *LandF* and *StateK* still emerge with a positive value. Compared with Model (2), R^2 increases by only 0.001 (0.968–0.967) even when many other variables such as *Urban*, *Popu*, *Complet*, etc. are taken into consideration. This result indicates that collusion is the culprit of exorbitant house prices and that other factors on house prices could be neglected from the perspective of statistics. Model (5) shows that an increase of 1% in *LandF* pushes up house prices by 0.211% and that an increase of 1% in *StateK* pushes up house prices by 0.049%.[15]

The significance level of *EconH* in Model (4) is relatively lower that of *EconH* in Model (3), but it is not significant in Model (5). This regression result shows *EconH* is not a robust indicator on house prices. It may be explained by the various irregularities in the distribution and purchase of economical houses. According to the state law, only those families with medium and low income have the right to purchase substandard economical houses and any applicant should submit an application for strict examination and approval by some departments concerned. Yet, poor management multiplies loophole opportunities for crafty officials. First, officials relax the restrictions in economical house approving and obscure the approval procedures to the extent that many wealthy people displace low-wage people to become the proprietors of economical houses. Many residential quarters intended for economical house proprietors are paradoxically flooded with luxury cars and occupied by some decently furnished yet idle houses; meanwhile, reselling and rental rates of economical houses are very high.[16] Second, due to lenient construction requirements in the early years, economical houses failed to reflect its characteristics of being economical. Third, the beneficiaries of economical houses are only limited to local urban residents and a lot of migrant workers from other regions are not entitled to housing benefits. Due to mismanagement, economical houses cannot give a full play to its role in curbing house prices.

The effects of other factors on house prices consist with our theoretical prediction on the whole. First, from the perspective of demand, the variable *Urban* at the significance level above 1% significantly increases house prices. This discovery is in accord with Zou and Niu (2010) and Xu et.al. (2012). *Popu* is not a robustness indicator on house prices either, which shows that population structure cannot effectively affect house prices even if the factor collusion is controlled. Chen and Fu (2011) made a similar discovery in their research without taking collusion into account. Second, from the perspective of supply, the coefficient of *Complete* is not significant although its minus value consists of theoretical prediction. The variable *Cost* reflects construction cost and its significantly positive coefficient shows it is also in accord with theoretical prediction: the higher the cost of completed houses, the higher the house prices. Zou and Niu (2010), Chen and Fu (2011) and Lu (2011) drew the same conclusion in their respective research. The significantly positive coefficient of *Rate* seems to be at variance with theoretical prediction, but as Xu (2011) asserted, the relatively low nominative interest rate perversely turns out to be a catalyst for a great demand for investment property.

4.5.4 *Robustness Test*

In our analysis, the incentive for the local government to collude with the real estate enterprise is to obtain land transaction fees so as to ease the pressure of imbalance of fiscal expenditure and fiscal revenue. Therefore, a potential problem is that collusion apparently exists only in underdeveloped regions whose fiscal revenue is insufficient, whereas in developed regions it seems unnecessary for the local government to collude with the real estate enterprise because house prices are determined purely by supply and demand. To remove the doubt, we classify the 31 provincial-level administrative units that the sample covers into different groups in terms of the degree of fiscal pressure, i.e., the ratio between fiscal expenditure and fiscal revenue. For each of the 31 provincial-level administrative units, we computed the average of the fiscal pressure during the 11 years between 2000 and 2010. Then, we divide the 31 provincial-level administrative units into "high fiscal pressure group" (above the median) and "low fiscal pressure group" (below the median). The median is 2.15, representing the ratio between fiscal expenditure and fiscal revenue. "High fiscal pressure group" includes Jilin, Anhui, Jiangxi, Hebei, Hunan, Guangxi, Sichuan, Guizhou, Yunnan, Tibet, Shaanxi, Gansu, Qinghai, Ningxia, and Xinjiang (Sinkiang). "Low fiscal pressure group" includes Beijing, Tianjin, Hebei, Shanxi, Inner Mongolia, Liaoning, Heilongjiang, Shanghai, Jiangsu, Zhejiang, Fujian, Shandong, Henan, Guangdong, Hainan, and Chongqing. By using the quantitative model mentioned previously, we obtained the following regression results of the two subsample data (See Table 4.6).

Table 4.6 Subsample regression results of housing variables

Explanatory variables	*Dependent variable: Selling price of commercial residential housing*			
	High fiscal pressure group		*Low fiscal pressure group*	
	FE (1)	*OLS (2)*	*FE (3)*	*OLS (4)*
LandF	0.474***	0.465***	0.138Ψ	0.368***
	(0.189)	(0.219)	(0.094)	(0.094)
StateK	0.042***	0.017***	0.058***	0.131***
	(0.019)	(0.020)	(0.022)	(0.016)
EconH	0.015	−0.031	−0.006***	−0.022
	(0.018)	(0.025)	(0.023)	(0.031)
Urban	0.131	0.264***	0.355	1.379***
	(0.116)	(0.083)	(0.339)	(0.191)
Popu	0.006	0.002	0.006*	−0.002
	(0.004)	(0.002)	(0.004)	(0.004)
Complete	−0.079	0.028	−0.006	−0.092*
	(0.048)	(0.033)	(0.050)	(0.028)
Cost	0.355***	0.475***	−0.044	0.639***
	(0.077)	(0.086)	(0.105)	(0.081)
Rate	0.375***	0.297***	0.545***	0.080**
	(0.065)	(0.051)	(0.072)	(0.050)
Region	Y		Y	
Year	Y	Y	Y	Y
R^2	0.938	0.891	0.969	0.932
Observations	129	129	130	130

Notes Robustness standard error is employed for all regression models. All values in the parentheses denote a standard error. The signs ***, **, *, and Ψ denote the significance level of 1, 5, 10, and 15% respectively

4.6 Summary

To explain why house prices in China are so exorbitant, we introduce the behaviors of the central and local governments into the conventional theoretical framework of demand and supply and construct a new convincing neo-political economics framework. We argue that, against the background of political centralization and fiscal decentralization, the local government colludes with the real estate enterprise so as to push up house prices and GDP aggregate and gain political promotion and fiscal revenue by controlling land supply. We chose land transaction fees and economic housing investment to indicate the level of collusion and discovered that collusion significantly increases house prices. In addition, construction cost and urbanization and many other factors also affect house prices. The policy implications of this chapter are as follows:

Fundamentally, China needs an institutional arrangement of moderating house prices which should readjust the fiscal decentralization between the central and local governments, grant the local government more fiscal autonomy or ease its financial burden so as to weaken its dependency on land finance, and eventually undermine the incentives of collusion.

Second, for economical houses to make their full play in curbing house prices, the central government must see to it that improvements are made on construction requirements and allocation modes and that only those medium- and low-wage people are the beneficiaries of economical houses.

Third, the central government should forbid state capital (especially the capital of state-owned enterprises) to flow into the real estate market, cutting off the direct link of interest conveyance between local governments and real estate enterprises.

Fourth, the central government should liberalize the supply of land market, allow collective land to enter into the secondary market, and eliminate the local government's monopolistic power over land supply.

Notes

1. The three big mountains refer to imperialism, feudalism and bureaucrat-capitalism that oppressed the Chinese people before the Communist Party of China founded the People's Republic of China in 1949. The contemporary Chinese media and communities satirically coined the

concept of the new three big mountains—exorbitant house prices, high tuition fee and staggering medical charges—to show the poignant life burden that a contemporary Chinese has to bear.

2. See http://news.xinhuanet.com/fortune/2011-02/10/c_121059274.htm (Accessed August 8, 2011).
3. See the website of the Bureau of Beijing Statistics. http://www.bjstats.gov.cn/sjjd/jjxs/201105/t20110506-201587.htm (Accessed February 13, 2012).
4. The calculating tool of housing loan is provided by soufun.com, the most influential housing website. See http://soufun.com/house/tools.htm. This tool is available at any time so long as we click and open the website.
5. Xu (2006) observed that a large proportion of fat profits from exorbitant house prices was seized by some local governmental departments and officials and thus aggravated corruption. Chen and Gao (2012) argued that income gap and house prices would be mutually intensified, i.e. exorbitant house prices would restrain the consumption of the poor and further widen income gap. Early on, Rodda (1994) made a research on the relationship between house prices and income gap.
6. Here encouragements and restraints respectively corresponds to aggressive strategy and conservative strategy discussed in Chap. 2.
7. See "State Council's Special Inspection of Execution of Real Estate Market Regulations", website link: http://www.gov.cn/jrzg/2012-07/24/content_2190967.htm (Accessed July 25, 2012).
8. The term "the local government" is an abstract subject, but the real subject is local officials. Under China's cadre personnel system, chief officials in the local governmental departments are highly mobile from one city to another or from one position to another.
9. Take the notorious event of "rent replacing requisitioning" in Xianghe County, Hebei Province. In 2008, the Xianghe county government (Magistrate Zhang Guijin) orchestrated a farmland grab, signing an agreement with some village or township party committees that land was requisitioned at an annual rent of CNY1150 per μ (a Chinese measurement unit of land area, equal to 1/15 hectare). However, a year later, the county government auctioned off the requisitioned land to real estate enterprises at CNY 600,000 per μ, 520 times more than that the requisition price. According to media reports, the case involves more than 4000 μ (about 266.7 hectares) of farmland. (See "Illegally Enclose Land at Low Price and Resell it to Real Estate Developers at Marked-up Price in Xianghe County, Hebei Province"), http://society.people.com.cn/GB/14675006. See also http://en.people.cn/90001/90776/90882/7390106.html (Accessed May 16, 2011).

10. See "A National Survey of Policies of Readjusting and Controlling Real Estate Markets in China", http://money.msn.cn/zhuanti/gdfdctkmb (Accessed July 8, 2011).
11. See "Why Is Readjustment and Control of Real Estate Is Brushed off by Local Governments?" in *China Economic Weekly*, 28 August 2006.
12. See "Slow Progression of Constructing Indemnificatory Houses in China", xinhuanet.com, October 28, 2009. http://news.xinhuanet.com/politics/2009-10/28/content_12350785.htm (Accessed July 8, 2011).
13. "Greentown President Song Suspected of Greasing Palm of Party Secretary of Zhoushan", *Yangtze Evening Post*, January 18, 2012.
14. The variable *IndemH* (denoting indemnificatory housing investment) should be a better substitute for *EconH*, but the State Statistics Bureau just provided information on *EconH* and the data of total *IndemH* is inaccessible.
15. Zhou Bin and Du Liangsheng (2010) also discovered land fiscal revenue was the reason for skyrocketing house prices. Nevertheless, their research lacked a micro-base on which to explain the local governments behaviours and the central government's periodical regulation and readjustment of house prices.
16. For more information, see a series of reports on economical houses in sohu website pages for financial news and analysis. http://business.sohu.com/s2009/jjsyf/ (Accessed August 16, 2011).

References

Bo, Zhiyue. "Economic Performance and Political Mobility: Chinese Provincial Leaders". *Journal of Contemporary China*. 1996, 5 (12): 135–154.

Carstensen, Kai, Oliver Lsewig, and Timo Wollmershauser. "Monetary Transmission and House Prices: European Cross-country Evidence". CESifo Working paper, 2009.

Chen, Chen, and Yong Fu. The Decision of High House Prices in China: Bubbles, Fundamentals and Periodical Fluctuations — An Empirical Research Based on Panel Data (1999–2009). Working paper, 2011. 陈晨, 傅勇. 中国高房价的决定: 泡沫、基本面与周期波动—基于板面数据的实证研究 (1999–2009). 工作论文, 2011.

Han, Li, James Kai-sing Kung. "Fiscal Incentives and Policy Choices of Local Governments: Evidence from China". Working paper, 2012.

Jin, Hehui, Yingyi Qian and Barry Weingast. "Regional Decentralization and Fiscal Incentives: Federalism, Chinese Style." *Journal of Public Economics*. 2005, 89 (9–10): 1719–1742.

Kuang, Weida. "Expectations, Speculative Binges and Fluctuations of House Prices in China's Cities". *Economic Research Journal*. 2010(9): 67–78. 况伟大. 预期、投机与中国城市房价波动. 经济研究. 2010(9): 67–78.

Li, Hongbin and Li-An Zhou. "Political Turnover and Economic Performance: The Incentive Role of Personnel Control in China." *Journal of Public Economics*. 2005, 89: 1743–1762.

Li, Yonggang, and Xiang Li. Fiscal Decentralization, Local Governmental Competition and House Prices Fluctuation: An Empirical Research on China's 35 Large- and Medium-Sized Cities. *Soft Sciences*. 2012, 26(1): 42–46. 李勇刚, 李祥. 财政分权、地方政府竞争与房价波动: 中国35个大中城市的实证研究. 软科学. 2012, 26(1): 42–46.

Liang, Yunfang, and Tiemei Gao. "An Empirical Analysis of the Regional Differences of Real Estate Price Fluctuations in China". *Economic Research Journal*. 2007(8): 133–142. 梁云芳, 高铁梅. 中国房地产价格波动区域差异的实证分析. 经济研究. 2007(8): 133–142.

Lü, Bingyang and Huihua Nie. Flexible Sharing Contracts: The Contract and Economic Performance of China's Tax-sharing System. Economic Theory and Business Management. 2014(7): 43–50. 吕冰洋、聂辉华. 弹性分成:分税制的契约与影响. 经济理论与经济管理. 2014(7)：43–50.

Lu, Chiwen. A Research on Effective Strategy of Curbing Rapidly-Increasing House Prices. Public Finance Research. 2011(10): 58–60. 卢驰文. 遏制房价过快上涨的有效策略研究. 财政研究. 2011(10): 58–60.

Mankiw, Gregory, and David Weil. "The Baby Boom, the Baby Bust, and the Housing Market". *Regional Science and Urban Economics*. 1989, 19 (2): 235–258.

Nie, Huihua, and Minjie Jiang. Collusion and Coalmine Accidents: Evidence from Provincial- Level Panel Data. *Economic Research Journal*. 2011(6): 146–156. 政企合谋与矿难：来自中国省级面板数据的证据. 经济研究. 2011(6): 146–156.

Nie, Huihua and Qiaochu Li. "House Purchase Restriction Is Hard to Curb the Increasing Trend of House Prices". *Enterprise & Financing*, Supplement to *The Journal of Finance and Management Research*. 2011 (7). http://finance.ifeng.com/a/20110901/4513596_0.shtml. (Accessed June 15, 2012) 聂辉华, 李翘楚. 限购令难以遏制房价涨势. 企业理财. 金融与管理研究增刊. 2011 (7).

Nie, Huihua and Jinbo Li. Collusion and Economic Growth. *China Economic Quarterly*. 2006, 6 (1): 75–90. 聂辉华, 李金波. 政企合谋与经济发展. 经济学(季刊). 2006, 6(1): 75–90.

Nie, Huihua and Jinbo Li. "Collusion and Growth: A New Perspective on China Model". *Economic and Political Studies*. 2013, 1 (2): 18–39.

Qian, Yingyi and Chenggang Xu. "Why China's Economic Reforms Differ: the M-form Hierarchy and Entry/expansion of the Non-state Sector". *Economic Transition*. 1993, 1 (2): 135–170.

Rodda, David. *Rich Man, Poor Renter: A Study of the Relationship Between the Income Distribution and Low Cost Rental Housing*. Harvard University, doctoral dissertation, 1994.

Wu, Jinhua. Deep Roots for the Rise of House Prices in Urban China and Solutions Thereof. *Economic Management Journal*. 2010(4): 167–172. 我国城市房价上涨的深层原因与对策. 经济管理. 2010(4): 167–172.

Xu, Jianguo. Increased House Prices Driven by Low Interest Rates: Evidence from China, U.S., and Japan. 2011(12): 5–13. *Shanghai Finance*. 许建国. 低利率推高房价: 来自中国、美国和日本的证据. 上海金融. 2011(12): 5–13.

Zhang, Yanjun, Yuanzhi Hou and Wenlong Du. The Impact of Land System and 'Land Finance' on Rising House Prices and Strategies of Curbing it. *Journal of Socialist Theory Guide*. 2012 (2): 16–18. 张燕君, 侯远志, 杜文龙. 土地制度与 "土地财政" 对房价上涨的影响及平抑对策. 理论导刊. 2012 (2): 16–18.

Zhou, Bin, and Liangsheng Du. Land Finance and Rise of House Prices: A Theoretical Analysis and Empirical Research. *Finance & Trade Economics*. 2010(8): 109–116. 周彬, 杜两省. 土地财政与房地产价格上涨: 理论分析与实证研究. 财贸经济. 2010(8): 109–116.

Zou, Zhizhuang, and Linlin Niu. "Demand and Supply for Residential Housing in Urban China". *Journal of Financial Research*. 2010(1): 1–11. 邹至庄, 牛霖琳. 中国城镇居民住房的需求与供给. 金融研究. 2010 (1): 1–11.

CHAPTER 5

Pollution, Illegal Land Use, and Other Major Social Problems

This chapter attempts to give an interpretation of pollution and illegal use of land and other major social problems within the principal-supervisor-agent analytical framework devised in Chap. 2. Section 5.1 explains environmental pollution from the perspective of collusion. Sections from 5.2 to 5.6, respectively, analyze the impact of collusion on illegal use of land, local negative externalities, smuggling, food safety, jerry-built projects, and adulteration. Section 5.7 is a summary.

5.1 Environment Pollution

5.1.1 An Overview of Pollution in China

To be frank, China is one of the most polluted countries all over the world. In 2006, New York-based Blacksmith Institute listed Linfen, Shanxi Province and Tianying, Anhui Province in its annual "10 worst" report.[1] Linfen, this soot-blackened city in the heartland of China's coal belt makes Dickensian London look as pristine as a nature park. In China, air pollution, water pollution, and solid waste pollution are harrowing. According to a report released by the official website of the 2008 Summer Olympic Games, "A conservative estimate by Chinese experts shows that the economic losses incurred by environmental pollution and ecological destruction amounted to as high as CNY200 billion, accounting for nearly 9% of China's gross national product (GNP)."[2] The main points of the report are as follows:

H. Nie, *Collusion, Local Governments and Development in China*,
DOI 10.1007/978-981-10-5059-6_5

1. Pollution in cities and townships is extremely severe. In China, air in urban areas is predominantly polluted by smoke, soot, dust, and acid rain (sulfur dioxide, SO_2), among which pollution due to dust and acid rain is most harmful. The Global Atmosphere Watch (GAW) monitoring results show that in Beijing, Shenyang, Xi'an, Shanghai, and Guangzhou, daily average concentration of total suspended particulates in the atmosphere ranged from 200 to 500 micrograms per cubic meter, 3–9 times higher than the World Health Organization standards. These five cities were all listed among Top 10 polluted cities. Beijing is one of the most polluted state capitals of the world. In addition, South and Southwest Acid Rain Control Zone, covering Guangdong and Guangxi, Guizhou and Sichuan Basin has been listed in the three severest acid rain zones. As to the water pollution, nearly half of river stretches of all the seven drainage systems (water systems) have been polluted. The concentration of harmful and toxic substances in 86% of river stretches flowing through cities exceeds the permissible limits. In atmospheric sounding and noise pollution (disturbance), ambient noise level in urban thoroughfares and roads is more often than not between 70 and 76 decibels, and 98.2% of urban noise levels are above the permissible limit of 70 decibels. Two-thirds of urban residents are exposed to hazardous noise levels. Stockpiles of industrial solid waste increase by 20 million tons annually on average.
2. Environmental pollution has spread from urban to rural areas. In China, township and village enterprises have polluted two-thirds of rivers and more than 1.3 million hectares of land. Every year, agricultural chemical fertilizers are heavily applied to approximately 160 million hectares of land, but effective the rate of fertilizer application is as low as 30%, two times less efficient than advanced agricultural areas in developed countries. The other 70% of agricultural chemicals evaporate into the atmosphere or dissolve into soil and rivers and lakes, causing eutrophication of waters in some regions and excessive nitrates in drinking water.
3. Ecological destruction is increasingly severe. In China, about 4600 species (or 15%) of higher plants are being endangered or on the verge of extinction. Over the past five decades, about 200 species of higher plants have become extinct, averaging 4 every year. About 400 species of wild animals are being endangered and some

of them on the verge of extinction too. Large tracts of timberland and wetland have been encroached. Grassland is being severely degenerated. China is also severely impacted by land desertification which encompasses over 30% of its territory. From 1950s to 1970s, an annual average of 600 square miles of land was desertified; from 1970s to 1980s, the annual average increased to 810 square miles. The total desertified land amounted to 77,606 square miles. Soil erosion by water is also a severe problem in China. At present, the area suffering from soil erosion is nearly 700,000 square miles. Over the past decades, due to overgrazing and mismanagement, over 330,000 square miles of grassland have been degenerated, desertified, or salinized, aggravating soil erosion and wind-sand damage.

5.1.2 Collusion and Pollution

According to the analytical framework of collusion, environment pollution is an inevitable consequence. In order to develop the local economy, the central government introduces either highly-polluting and low-cost industries or low-polluting and high-cost industries. A lot of manufacturing industries such as cell and battery factories, plastic and chemical plants, paper mills and tanneries are all highly-polluting, and only when large amounts of funds are invested in these industries for technological innovation can pollution be reduced. However, in the short term, technological innovation increases production cost, reduces profits, and slows economic growth. If the central government or an upper administrative organ thinks local economic growth outweighs social stability, or accepts the concept of "developing economy and protecting environment in the same time", it will connive at the introduction of highly-polluting and low-cost industries. Thus, high-speed economic growth is concomitant with severe environment pollution.

In environmental protection, there is a law known as Environmental Kuznets Curve (EKC) which shows an inverted U in the relationship between environment destruction and per capita income (Shafik 1994; Grossman and Krueger 1995). To be specific, the lower the economic development level, the less industrial output and the less environment pollution; the higher the economic development level, the more industrial output and the severe environment pollution; the more advanced the economy, the more importance attached to environment protection and the more resourceful and capable to curb environment pollution. Though EKC's application is contestable, some experts (Jiang et al. 2011) tested it with data collected from China. Interestingly, the government-firm collusion model can provide a microscopic basis for EKC. According to the collusion model, in the primary stage of development when material life is scarce, the central government allows collusion because it believes that economic growth outweighs social stability; on the contrary, in the advanced level of development when residents care more about environmental quality and protests against environmental pollution, the central government will prevent collusion and demand the local government and the enterprise to choose a "good" mode of production. This logical reasoning is applicable not only to the explanation of China's pollution governance but also to that of other developing countries.

China is currently in its intermediate stage of development, so economic growth is placed above social stability. In most regions, local governments still insist on the concept of "prioritizing economic development over environment protection". According to our collusion model, high economic growth should be concomitant with severe pollution. Therefore, the case should be like this: in the area where the economy is accelerating, environmental pollution should be much severe. Coincidentally, the above assumption is proved. Map 5.1 shows the distribution of pollution sources of enterprises under the key monitoring of the state. Geographically, most polluting enterprises are distributed in the southeastern coastal areas and North China, where economy develops at a relatively faster speed.

5.2 Illegal Use of Land

With the development of the economy, factors of production such as nonrenewable resources whose elasticity of supply is nearly zero will become more precious. There is no exception to land. In order to guarantee sustainable supply of land and protect "the red line of 120 million hectares of arable land for food security," the Chinese central government strictly controls land use. The State Council and the Ministry of Land and Resources often jointly send inspection teams to provinces, municipalities, and autonomous regions to monitor the execution of *Land Administration Law*. Even so, illegal use of land is still rampant throughout the country. Illegal use of land refers to such a series

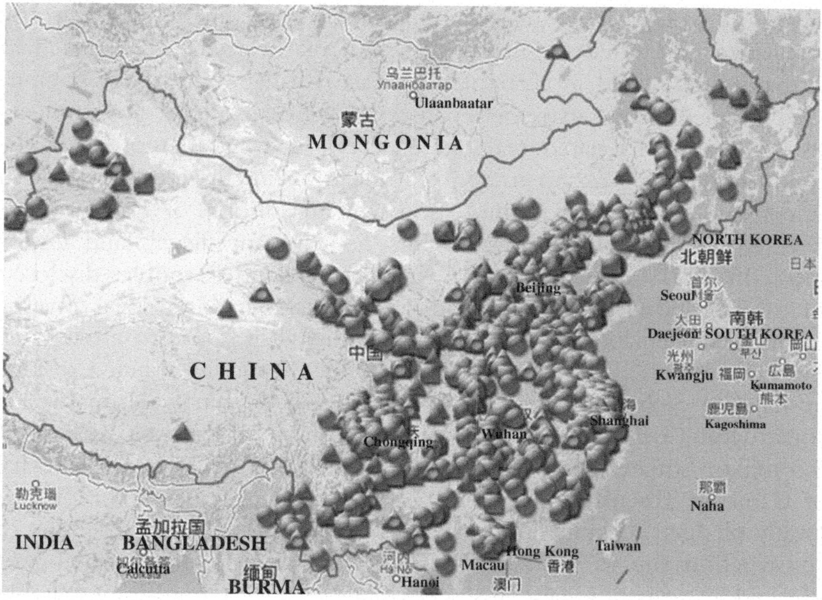

Map 5.1 Distribution of pollution sources of enterprises under key monitoring. *Note* This map is drawn by the author of the book according to the Institute of Public and Environmental Affairs (IPE), link: http://www.ipe.org.cn/. Different signs show different types of pollution source: stands for waste gas and wastewater, for waste gas; for wastewater; and for sewage disposal works

of behaviors as illegally reselling, seizing, damaging arable land and approving applications for land use. According to Zhang Xinbao, chief of the Enforcement Supervision Bureau of the Ministry of Land and Resources, during the first 7 years (1999–2005) after the issue of *Land Administration Law*, altogether more than 1 million cases of illegal use of land have happened, involving nearly 330,000 hectares of land, almost 70,000 hectares more than the total amount of newly-added land for construction in 2004. Of all the cases involving use of newly-added land for construction, cases of illegal use accounted for 34% annually on the average, the figure being as high as 80% in some provinces.

The subject of illegal use of land is local governments. According to statistics by Liang (2009), between 1999 and 2005, in all the cases of illegal use of land, the land area which local governments were directly implicated accounted for nearly one-third of the total; if calculated in terms of arable land, it accounted for nearly a half. The above computational results were extremely conservative because local governments also indirectly instigated real estate enterprise to grab more land for commercial purposes. So to speak, local governments are the de facto land monopolists. That is the reason why an editorial of *People's Daily* directed attack at local governments for rampant illegal land enclosure.[3] In 2001, a bulletin by the Ministry of Land Resources disclosed 11 cases of illegal use of land, and 7 of the cases involved local governments which enclosed arable land for industrial purpose to attract foreign investment.[4] For instance, in July 2000, Guangtai Ceramic Co. Ltd. in Jiajiang, Sichuan applied with Ganjiang Township government for constructing a floor tile factory with an annual production capacity of 3.5 million square meters of floor tiles. Luo Jiangquan, Party Secretary of Ganjiang Township, Mayor Li Xueyuan and Vice-mayor Li Zhiguo jointly pledged to grant a permit and allowed the investor to choose a construction site as it wished. Under the support and protection of Mayor Li Xueyuan, Guangtai signed an agreement, requisitioning arable land of about 2.57 hectares.

Why are local governments the backstage driving force of illegal use of land? A reference to the analysis of real estate markets in Chap. 4 is conducive to our understanding of the question. By enclosing arable land for industrial or commercial use, officials of local governments can obtain a lot of land transaction fees within their short tenure and meanwhile propel the development of local industries, gaining fiscal revenue and political capital. Liang (2009), by analyzing the data of 30 provinces, municipalities, and autonomous regions (excluding Tibet) in mainland China, measured the degree of illegal use of land with the land area involved in illegal cases and discovered that the higher the economic level, the higher the degree of fiscal decentralization, or the higher the land transaction fees, the severe the degree of illegal use of land. On the other hand, industrial and real estate enterprises purchased large tracts of land for a song, made substantial profits, and even could evade legal punishment under the "protective umbrella" of local governments. Since the annual quota of land approved for construction is very limited, local governments vie with peers—i.e., with local governments at the same administrative level—for more land in the stiff regional competition under the camouflage of attracting and inviting investments, which is an assessment criterion of political performance.

Owing to insufficient quota of land for commercial development, local governments also pull down residents' houses and ancient buildings of historical interests, seizes peasants' homestead land, and even transforms urban green land into industrial or commercial land. Illegal enclosure of land jeopardizes food security, harms peasants' interests, and destroys natural and cultural environments. Zhang Li et al. (2011) revealed the effect of collusion on illegal use of land. Basing their research on the 1999–2007 data of 30 provinces, municipalities, and autonomous regions (excluding Tibet), they measured the degree of illegal use of land with the following parameters: area of land and arable area illegally encroached and the number of illegal cases. They also noticed the effect of local promotion (dummy variable) on the level of collusion.[5] They discovered that, if compared with that in those provinces where governors are transferred from other provinces or appointed by the central government, the area of illegally encroached land in the provinces which are locally promoted increased by 30% on the average, the area of illegally encroached arable land increased by more than 50%, but the number of illegal cases that eventually received investigation and punishment decreased by 15%. These above figures show that collusion aggravates local governments' illegal use land and that it is more difficult to investigate and punish illegal cases.

5.3 Local Negative Externalities

In the analytical framework, we assume that the incentive for the local government to collude with the real estate enterprise is fiscal and political gains accompanied with economic growth. Fiscal and political gains are associated with regional competition because the central government assesses local officials through horizontally intergovernmental competition, i.e., competition between local governments at the same administrative level. Intrinsically, competition is conducive to economical and social development. Theoretically, the central government could either formulate policies to encourage competition in the aspects of public services, environmental quality, and anticorruption or replace GDP with other measures supportive of sustainable development such as Human Development Index (HDI) and Green GDP. Therefore, the main problems consist in asymmetry of information and regional differences of development cost resulting from the problematic assessment criterion of economic growth; however, these are the two problems that create loophole opportunities for collusion between local governments and local enterprises.

However, situations might go even worse than the decline of social well-being in a particular province or a county where collusion and regional competition co-exist because it will unavoidably result in local negative externalities and further aggravates the social welfare of the whole country at large. Suppose in provinces A and B where both the provincial governments collude with local enterprises. Seen from the perspective of locality, collusion just merely seems to do harm to the long-term interests of the local people. Now, let us consider the interactive behaviors between the two provincial governments. Reasoning from the sound premise that regional competition determines local fiscal revenue and local officials' political promotion, we could naturally draw a conclusion that the "reasonable" choice for each region is the adoption of a "bad" mode of production, such as introducing polluting industries and furthermore transferring the cost of the bad mode to other especially neighboring regions. Therefore, each region not only gains growth but also increases the cost of growth in other regions. As a matter of fact, this is a prisoner's dilemma, which is intrinsically a beggar-thy-neighbor mode of growth. Oates and Schwab (1988) discovered the negative externalities of local competition, i.e., likely transference of pollution due to competition.

Interesting, yet, seemingly unsolvable, such things frequently happen in real life. The magazine *China Environment Watch* published a pollution case in which three knockout enterprises got involved, disclosing a series of ruinous results due to collusion and regional competition.[6]

In 2004, the Taishun County government in Zhejiang Province began to invite investments. Village Shuiwei in Township Moon Lake (administratively under the jurisdiction of Taishun County, Zhejiang Province) became the most ideal construction site for polluting factories because this southernmost village in Zhejiang Province is adjacent to Fujian Province. Three enterprises which had actually earlier been eliminated by Wenzhou City began their construction here in succession. On the east of this village are Zhongchen Stainless Manufacturing Co. Ltd. and Kada Valve Casting Co. Ltd., with a county-level highway no less than five meters wide separating them; on the east of the village is Taigang Stainless Steel Group Co. Ltd.

Since then, smoke and soot discharged by the three enterprises threatened villagers' life. Blower fans rumbled day and night; black, acrid, and choky smoke gushed from chimneys of factories, enshrouding the whole village. Even if villagers shut their doors and windows tightly, they could not resist the odorous smell. Living under so terrible an environment for 8 years, some villagers felt debilitated, queasy, and affected with cough and other respiratory diseases. Mr. Xie, a villager examined abnormal in his lungs in a physical examination, complained, "Our village is surrounded by 'life killer', both water and soil have been seriously contaminated. It is not livable at all."

Serious pollution in this village was a self-evident fact, but officials of the Taishun County Environment Protection Administration brazenly claimed "a close and through examination" showed that pollutants discharged from the three enterprises were all up to the standard level. They even recklessly made it known that the three knockout enterprises had been actively engaged in "rectifying and reforming" antipollution mode of production. This flagrant defiance against the living condition of the local people was thoroughly ignored by the county government. More preposterously, the Taishun County government even had the imprudence to hold a preparatory meeting, intending to apply for "the first ecological county in Zhejiang province." Repeatedly urged by Shuiwei villagers, Taishun County Environment Protection Administration agreed to inspect the enterprises. However, each time when officials came for inspection, the three enterprises would stop their production and officials would draw the same conclusion that no pollution existed in the village. "How could they say there is no pollution in our village? Obviously, officials themselves had tipped the enterprises off their inspection work. Before the inspection team could come, factories had closed temporarily," grumbled the villagers angrily. "They pay no heed to our health. Ashes and dust could even fly into our bowls when we eat. We could not even have a wink all night. We breathe dirty air every day. How dare they say so?"

5.4 Smuggling

Smuggling that implicates in governmental officials is another form of collusion. By conniving at or directly getting involved in such an illegal activity, local officials can get bribery or perks from enterprises. Some governmental officials are even in cahoots with customs officers, helping local enterprises to evade tariffs. This *tacit quid pro quo* actually increases the financial strength of local enterprises so that they will be in a more favorable position to invite foreign investment than their law-abiding rivals in other regions (Cai and Treisman 2004). However, this illicit competition has a seriously negative impact on incentives for choosing a good yet costly mode of production. Such being the case, the rivals have no alternatives but to exploit a similar mode. In the collusion model, the central government becomes the victim because it suffers a great loss of tax revenue. Obviously, compared with tax payment and investment inviting in accordance with the law, smuggling is an extremely bad mode of production.

Take, for instance, the Zhanjiang smuggling case.[7] In the 1990s, smuggling was rampant in Zhanjiang, Guangdong. Especially after 1996, smuggling increased in an almost uncontrollable way in this haven for smugglers. Of the three most influential smuggling groups, a ring-leader nicknamed "Lord of Car Smuggling" illegally transported 1905 cars and 10,000 tons of diesel oil into domestic markets and evaded customs duties worth of CNY170 million in less than 2 years from May 1996 to April 1998 under the "protective umbrella" of his father Chen Tongqing, the then City Party Secretary of Zhanjiang. In order to find a market for smuggled cars, this city party secretary established a powerful network of *guanxi*, a relationship that defines an individual's place in the social structure and providing security, trust, and prescribed roles; in order to make his son awash with funds, he exerted influence upon the Zhanjiang City Government which later guaranteed a loan of HK$ 40 million for his son from a foreign bank. In addition, the Party Secretary also got involved in another smuggling group which ganged up with Chikan District Anti-Smuggling Office and Frontier Inspection Station to set up an auction for smuggled commodities. In this astonishing case, a batch of governmental officials including Cao Xiukang, Zhanjiang Commissioner of Customs and Excise, Zhu Xiangcheng, Head of Investigation Department, and Fang Peng, Director of Port Customs Office and other more than 200 officers and law's minions and public servants in frontier inspection stations, public security, marine police, commodity inspection, harbor superintendence administration and shipping gave the green light to illicit smuggling, taking bribery valued at more than CNY7 million. It is estimated that total value involved in all smuggling cases in Zhanjing from 1996 to 1988 surpassed CNY10 billion, evading customs duties worth of CNY6 billion.

The most notorious smuggling case in coastal Southeast China is perhaps the Xiamen Yuanhua Smuggling Case. Lai Cangxing, president of Xiamen Yuanhua Group Co. Ltd. and principal criminal of the case, seduced a number of government officials down to cesspool to get a loophole for illicit smuggling. From 1999 to 2004, this group smuggled refined oil, vegetable oil, cars, and cigarettes with a worth of CNY53 billion approximately $6.4 billion in terms of the exchange rate at that time) and a customs-duty evasion up to CNY30 billion. The principal criminal of the case fled to Canada. In 2011, he was extradited back to China and was sentenced to life imprisonment for offering bribery and smuggling. Many high-ranking officials got involved in this case, including Li Jizhou, Vice-minister of the Ministry of Public Security, and Shi Zhaobin, City Party Vice-Secretary. Xiamen Yuanhua Smuggling Case is the biggest corruption and smuggling case in China ever since 1949. Local government officials realized that substantial investment objectively made by the smuggling group and the relatively cheaper smuggled goods and commodities (such as diesel oil and gasoline) were a lure to attract foreign investment, which they reckoned that would boom the local economy and soon bring them bumper fiscal revenue and more promising political career. As a matter of fact, the above-mentioned smuggling cases caused serious damage to the normal economic order and induced rampant corruption.

5.5 Food Safety

Food safety, an increasingly serious problem that threatens health, is also associated with collusion. By conniving at low-cost production mode that produces shoddy products, the local government intends to nurture more profitable enterprises, create job opportunities, and falsify political achievements. Local enterprises profiteer from these nasty production activities but many consumers get screwed. Food safety will undoubtedly increase the probability of the occurrence of large-scale mass incidents of civil disobedience.

Take, for instance, the infamous 2008 Sanlu Milk Powder Scandal.[8] On July 16, 2008, excessive melamine was found in milk and infant formula produced by Shijiazhuang-based Sanlu Group Co. Ltd., a joint venture which used to be the largest milk powder manufacturer in China. Its annual production of milk powder had ever ranked first in successive 15 years in China, taking up 18% of the market share.

Obviously, this famous joint venture had ever brought the local government a considerable sum of tax revenue and enviable political achievements. Therefore, even when many infants fed with this infant formula were reported to have died or to have been hospitalized, the local government still tried to cover up the truth so that it could continue its production. According to mass media, Sanlu Group submitted a written report about the contamination of milk powder to the Shijiazhuang City Government on August 2, 2008. After receiving the report, the city government immediately sent Zhao Xinchao, a vice mayor whose duty it was to be responsible for work safety and Zhao Wenfeng, Secretary-in-General of the municipal government, and a large band of officials from the Administration of Quality Supervision, Inspection and Quarantine (AQSIQ), the Food and Drug Administration, Industrial Promotion Bureau to Sanlu Group for inspection. Yet, the officials did not demand Sanlu Group to terminate production immediately but instead proposed to pacify victims' family members with hush money for fear that this incident might be alerted to higher authorities and the mass media. On August 29, Sanlu Group made a second written report to Shijiazhuang City Government, but it was not until September 8 that the Hebei Provincial Government received the report. This implies that the city government had been covering up the truth before the provincial government received it. The scandal raised concerns about food safety and political corruption in China.

5.6 Other Major Problems

In addition to coalmine accidents, environment pollution, real estate bubble, local negative externalities, smuggling and food safety, collusion also results in other social problems such as jerry-built projects, tax frauds, shoddy products, and resistance against the regulation and readjustment measures of the central government. We must make it clear that not all the above problems are inevitable outcomes of collusion but that collusion results in or worsens those problems.

Jerry-built projects can be found in many regions of China. The reason for their being widespread is that local government officials blindly construct image projects or achievement projects within their short tenure in the hope of being promoted after the expiration of the term of office. Leaders of local building companies embezzle the construction payment to grease the palms of governmental officials for receiving a quality acceptance certificate and lining their own pockets even if they build their project on the cheap and shorten the construction cycle.

Tax frauds are also associated with the protectionism of local governments. Under the condition of regional fierce competition, the key to attracting foreign investment is to provide a more preferential taxation policy. Therefore, some local governments reduce taxes or even exempt some enterprises from state taxes (not local taxes) in order to invite foreign investment. This is a race-to-the-bottom behavior. Through a quantitative analysis of the data from the Chinese Industrial Enterprise Database, Cai and Liu (2009) noted that tax evasion was in positive correlation to the degree of market competition of industries, i.e., the more competitive an industry is, the more serious tax frauds are. Fan and Tian (2012) made a research on a 2002 policy experiment that newly established enterprises should pay corporate or business income taxes to the State Administration of Taxation instead to a local taxation bureau. They noted that tax evasion in newly established enterprises which should pay income taxes to the State Administration of Taxation was much less serious than that in those old enterprises which paid income taxes to a local taxation bureau. This observation implies that local governments tend to help enterprises evade taxes in order to attract more investment in the cut-throat interregional competition.

Adulteration instigated by local governments is another form of collusion. A bad mode of production helps enterprises save cost and thus profiteer from it. Massive adulteration is usually regarded as a source of fiscal revenue and unemployment reduction. In the notorious Sanlu Milk Powder Scandal, adding excessive melamine is a form of adulteration. In addition to adulteration, piracy and trademark infringement also exist in many industries. Collusion is a catalyst for these illicit activities.

5.7 Summary

In this chapter, we explained, within the government-firm collusion mode, a variety of social problems including environment pollution, illegal use of land, local negative externalities, smuggling, food safety, jerry-built projects, adulteration, etc. Behind these serious social and economic problems is still the invisible institutional drive of collusion. So long as local governments have the incentive to make grandiose plans for unrealistic political achievements and the mode of production is not under an effective supervision system, they are probably to collude with enterprises and bring so much undesired attendant social cost.

The policy implications of this chapter are as follows. First, a thorough elimination of collusion entails a reform of the current performance assessment system for governmento fficials, stamping out pure GDP race and navigating local officials to lay more emphasis on people's livelihood and quality of economic growth. Second, it is necessary for the central government to establish and perfect an inspection tour system or set up inspection stations across the country to fight against the illegitimate behaviors of local governments and enterprises. The officials of the inspection teams should be frequently transferred from one team or place to another for fear that they could also be corrupted.

Notes

1. "Top 10 Most Polluted Places" *(Adobe Portable Document Format). Blacksmith Institute.*
2. See "Training volunteers about Green Olympiad: China's environmental problems", http://sports.sina.com.cn/s/2006-08-24/22342417025.shtml.
3. "How dare the local government take the lead in illegal use of land?" *People's Daily*, May 15, 2006.
4. The Ministry of Land and Resources, "Circular on the investigation and punishment of eleven cases of illegal uses of land", [2001] No. 175.
5. A provincial governor is promoted to his governorship through one of the following three ways: local promotion, transfer from another province and appointment by the central government. A provincial governor locally promoted is not necessarily a native but one who has served as a very influential role (such as a standing vice-governor) in the local government for quite some time. According to Zhang Li (2011), in the case of illegal use of land, collusion is more likely to happen when a provincial governor is locally promoted than the situation when a governor is a native. In testing coalmine accidents, Nie and Jiang (2011) perceived *native_g* (the vice-governor in charge of work safety) as a proxy variable.
6. For full text, see "Three knockouts accused of peccancy in Zhejiang due to mismanagement", http://www.gov-news.org.cn/2012/1022/27181/.html.
7. For full text, read "Intolerant of trading power for money—A full account of Zhanjiang smuggling case", *Guangming Daily*, June 8, 1999. *Guangming* literally means "throwing light on" in Chinese.

8. For full text, read "Governmental Officials prevent milk powder scandal spreading with hush money: A survey of the collapse of Sanlu". China Youth Daily, January 7, 2009.

References

Cai, Hongbin, and Daniel Triesman. "State Corroding Federalism". *Journal of Public Economics*. 2004, 88: 819–843.

Cai, Hongbin and Qiao Liu. "Competition and Corporate Tax Avoidance: Evidence from Chinese Industrial Firms". *Economic Journal*. 2009, 119: 764–795.

Fan Ziying and Tian Binbin. "Intergovernmental Horizontal Competition and Passive Corporate Tax Evasion: Evidence from Income Tax Sharing Reform". Working paper of Huazhong University of Science and Technology, 2012. 范子英, 田彬彬. 政府间横向竞争与企业被动逃税——来自所得税分享改革的证据。华中科技大学工作论文, 2012.

Grossman, Gene, Alan Krueger. "Economic Growth and the Environment". *Quarterly Journal of Economics*. 1995, 110 (2): 353–377.

Jiang, Yi, Tun Lin, and Juzhong Zhuang. "Environmental Kuznets Curves in the People's Republic of China: Turning Points and Regional Differences". SSRN working paper, 2011.

Liang Ruobing. Promotion Incentives, Departmental Interests and Illegal Use of Land under the Fiscal Decentralization. *China Economic Quarterly*. 2009 (1): 283–305. 刘若冰. 财政分权下的竞争激励、部门利益与土地违法. 经济学季刊. 2009 (1): 283–305.

Nie, Huihua, and Minjie Jiang. Collusion and Coalmine Accidents: Evidence from Provincial-Level Panel Data. *Economic Research Journal*. 2011 (6): 146–156. 政企合谋与矿难: 来自中国省级面板数据的证据. 经济研究. 2011 (6): 146–156.

Oates, Wallace and Robert Schwab. "Economic Competition Among Jurisdictions: Efficiency Enhancing or Distortion Inducing". *Journal of Public Economics*. 1988, 35: 333–354.

Shafik, Nemat. "Economic Development and Environmental Quality: An Econometric Analysis". *Oxford Econometric Papers*, 1994, 46: 757–773.

Zhang, Li, Xianxiang Xu, and Xianbin Wang. Local Collusion and Illegal Use of Land. *The Journal of World Economy*. 2011 (3): 72–88 地方官员合谋与土地违法. 世界经济.

PART II

The Impact of Collusion and Institutional Environment

CHAPTER 6

The Impact of Collusion on Corporate Behaviors and Political Performance Assessment

In this chapter and Chaps. 7 and 8, we will discuss various types of social cost (expressed as C, see the discussion of collusion model in Chap. 2) as a result of collusion. It is not that we deny possible gains, but that we feel it necessary to have a comprehensive understanding of negative costs of various forms of collusion. Undeniably, an analysis of collusion on corporate and governmental behaviors and on social development is conducive to helping the central government make an objective comparison between gains and costs.

This chapter mainly focuses on the impact of collusion on corporate behaviors and political performance assessment. By referring to the Chinese Industrial Enterprise Database, we use the proportion of state-owned shares in enterprises to measure the level of collusion, and use the capital input of R&D to measure technological innovation. The results show that collusion marginally reduces technological innovation. When using the proportion of subsidy income to sales volume to measure the level of collusion, we observe that collusion significantly reduces total factor productivity (TFP).[1] Sections 6.1–6.3 respectively analyze the impact of collusion on the capital input of R&D, TFP, and FDI.

H. Nie, *Collusion, Local Governments and Development in China*,
DOI 10.1007/978-981-10-5059-6_6

6.1 Collusion Hindering Technological Innovation

6.1.1 The Facts of Technological Innovation in Chinese Enterprises

Since Adam Smith, economists began an inquiry into the nature and causes of wealth of the nations. Contemporary economists have discovered that the wealth gap between nations arises from the TFP gap and that the TFP gap results from the gap of technological innovation (Caselli 2005; Hall and Jones 1999). Therefore, for China to gain more competitive edges in economic growth, its enterprises should be motivated to make innovation. As a rule of thumb, the indicators to measure the innovation level of an enterprise are R&D inputs and outputs. Since it is not easy to estimate such as indicators including the number of patents, the number of paper grants and citation index of scientific papers which are used to measure R&D outputs, R&D intensity—usually defined as the proportion of R&D investments to sales volume—has become a frequently used method for all countries to assess innovation. A higher R&D intensity index denotes that more emphasis has been laid on R&D and that it is more likely to achieve innovation. Table 6.1 lists the top-ranking Northern European and American multinational corporations (MNCs) in the respect of R&D investment in 2009. These MNCs make an annual investment of thousands of millions of Euros in R&D and the highest R&D intensity was 20%. Comparatively speaking, there was a large gap between Chinese enterprises and globally first-rate MNCs both in R&D investment and in R&D intensity. By using the Chinese Industrial Enterprise Database, we computed the 2001–2003 and 2005–2007 indices relevant to the measurement of R&D. Table 6.2 shows that R&D investment aggregate in the whole Chinese manufacturing amounted to only CNY 26.2 billion, much less than that of any of the top five MNCs in Table 6.1. Between 2001 and 2007, only 10% or so of Chinese enterprises made an investment in R&D. The average R&D intensity of Chinese enterprises was 0.17, 20 times less than the global average of 3.5%. The R&D intensity of the top percentile of Chinese enterprises in 2007 was only 5.87, much lower than that of the MNCs in Table 6.1.

6.1.2 Collusion Hindering Innovative Motivation

A Case Study: *The most scandalous event in antivirus software industry—Rising's false incrimination against Micropoint by Rising*

Table 6.1 Top 10 companies ranked by R&D in 2012

Rank	*Name*	*Country*	*Industrial sector*	*R&D (€ million)*	*R&D intensity (%)*	*Sales 2012 (€ million)*
1	Volkswagen	Germany	Automobiles & Parts	9515.0	4.9	193,000.0
2	Samsung	South Korea	Electronic & Electrical Equipment	8344.7	5.9	142,000.0
3	Microsoft	USA	Software & Computer Services	7890.7	13.4	59,003.3
4	Intel	USA	Technology Hardware &Equipment	7691.4	19.0	40,428.2
5	Toyota Motor	Japan	Automobiles & Parts	7070.9	3.7	193,000.0
6	Roche	Switzerland	Pharmaceuticals & Biotechnology	7007.8	18.6	37,622.3
7	Novartis	Switzerland	Pharmaceuticals & Biotechnology	6922.8	16.1	42,953.6
8	Merck US	USA	Pharmaceuticals & Biotechnology	5995.9	16.7	35,824.6
9	Johnson & Johnson	USA	Pharmaceuticals & Biotechnology	5809.5	11.4	50,950.4
10	Pfizer	USA	Pharmaceuticals & Biotechnology	5740.5	12.8	44,706.7

Source R&D ranking of the world top 2000 companies. http://iri.jrc.ec.europa.eu/scoreboard13.html (Accessed May 5, 2017)

The comparative analysis between Chinese enterprises and internationally well-known MNCs in Sect. (6.1.1) sparked our interest in the reasons why Chinese enterprises are reluctant to invest as much in R&D as global top-ranking MNCs? It is undoubted that only when R&D investments exceed potential earnings will an enterprise determine to give it up. Besides R&D and innovation, what other means could be employed to win in the increasingly fierce competition? A typical Chinese businessman or entrepreneur would immediately give a typical Chinese

Table 6.2 R&D investment of Chinese manufacturing sector

Year	*Number of enterprises*	*Percentage of enterprises engaged in R&D (%)*	*R&D inputs aggregate (million USD)*	*Average R&D intensity (%)*	*Top percentile R&D intensity (%)*
2001	164,587	12	26,292	0.14	2.99
2002	175,249	13	34,670	0.17	3.26
2003	191,415	13	43,202	0.16	3.43
2005	268,045	10	92,383	0.18	4.84
2006	297,630	10	124,000	0.20	5.46
2007	333,306	11	2843.0	0.19	5.87

Source All the figures are computed by the author of the book according to the data from the Chinese industrial enterprise database

answer: making *guanxi* transactions. Many business executives pursue such an unspoken rule or maxim that "it is only *guanxi* that is the primary productive force", a concept which in recent years has caught on among foreign firms with a long-standing business relationship with their Chinese counterparts. In all forms of *guanxi*, the one with local governments is undoubtedly the most important. Some Chinese entrepreneurs and businessmen would prefer to rub shoulders with governmental officials and corrupt them with bribery rather than make a risky yet rewarding investment in R&D and innovation.

In early 2010, Xinhua News Agency, *China Youth Daily* and *Economic Observer* and other mass media disclosed a scandalous collusion in which an antivirus software corporation, corrupting and, thus, shielded by a governmental network supervision agency, fabricated a series of charges against its potential competitor Micropoint. An account of the case is helpful for us to have a better understanding of how *guanxi* severely affects technological innovation in China.

In 2003, Liu Xu, initial designer of Rising antivirus software (RAV) and manager of Rising Technology Co. Ltd., resigned and founded Micropoint Information Technology Co. Ltd. In 2004, Tian Yakui resigned from Rising and joined in Micropoint. In 2005, the R&D team of Micropoint successfully innovated an antivirus software which could take active and initiative defense against viruses. Rising immediately felt Micropoint's innovative product would threaten their conventional products, which then accounted for almost 60% of antivirus software market share. The following is the process of Rising's collusion with a governmental agency responsible for network supervision:

- In June 2005, Rising began to corrupt Yu Bing, Director of Network Supervision Department of Beijing Municipal Public Security Bureau (BMPSB), offering him in succession a bribery of CNY 4.2 million in the hope of eliminating its competitor Micrpoint by fabricating evidence.
- On July 5, 2005, the BMPSB investigated Micropoint under the pretext of "examining the qualification of antivirus corporations" and moved a Micropoint computer storing its core technologies to the Rising's headquarters. Through law personnel in charge of the case, Yu demanded Liu that he should either sell its company to an established and government-backed corporation such as Rising or relocate its company to his birthplace Fujian.
- In August 2005, Yu prompted Beijing Simaite Management Consultant Firm and Beijing Jianqiao Securities Co. Ltd. to fabricate evidence that they both had a great financial loss. At the behest of Yu, an expert group was formed and they soon arrived at a conclusion that the loss was directly related with the infection of a virus spread from Tian's laptop computer.
- Later, Yu sent his men to force other three antivirus companies to frame up Micropoint by inventing false evidence of virus outbreak due to Micropoint's new antivirus program.
- In addition, Yu collaborated with Rising's Vice President Zhao Sizhang, instigating Beijing Zhongrunhua Accounting Firm to make a false assessment of the losses mentioned above and accused Micropoint of destroying computer information system and of infringing commercial secret.
- Yu-led Network Supervision Department fabricated and released a news report "The first case of spreading virus by a domestic antivirus firm" and sent it to Beijing TV Station in the name of the authorities for broadcasting nationwide.

It was not until July 2008 when the Beijing Municipal Commission for Discipline Inspection received a real-name tip-off about Yu's embezzlement that a special inspection group was formed to investigate the scandalous event. Though eventually Yu was handed a suspended death sentence and Micropoint obtained its sales license for initiative defensive antivirus software, it was stalled for more than 3 years and missed the golden opportunity to innovate.

This most scandalous event in antivirus software industry reflects, from a specific angle, the destructive effect of collusion on technological innovation of a local enterprise. In this game of collusion, Rising took the advantage of the opportunity of monopolizing profits and Yu took bribery and made "outstanding" political achievements in cracking down "illegitimate" corporations. In a particular case, the government is an impersonal entity and it is the officials who administrate the governmental departments and agencies that constitute the government. This is a typical collusion where many of the officials in the local government's departments or agencies are corrupted by an outdated corporation to suppress and bludgeon an innovative enterprise.

6.1.2.1 Theory: Collusion distorts incentives for development and research
The current academic research on corruption is mainly focused on political connections. In the economic and accounting academic communities, Fisman (2001), Faccio (2006), Fan et al. (2007), Claessens et al. (2008), Yu and Pan (2008), Wu et al. (2008), Luo and Tang (2009), and Chen et al. (2010) have analyzed the channels through which an enterprise establishes its political connections and pertinent impact thereof but none of them have discussed the relations between political connections and the innovation of an enterprise. Yang (2011), with the aid of a quality/vertical differentiation model, analyzes the problems of decision making which enterprises encounter when allocating resources to political connections and product R&D. He argues that an enterprise will be keen on political connections rather than capacity building if the government is granted large quantities of economic resources and great discretion on the one hand but it is very difficult for enterprises to upgrade and maintain the advantages of product quality on the other. He also notices that domestic enterprises are very likely to be willingly reduced to the producers of low-end services and products and to be more desirous of political connections if the income distribution system leads to a larger low-end consumer group and a larger gap between the poor and the rich, and meanwhile foreign MNCs occupy the market of high-end services or products.

The quality/vertical differentiation model is enlightening. Suppose an enterprise has two channels to make profits: one is a good mode of production, i.e., reducing cost through R&D investment and innovation and the other is a bad mode of production, i.e., corrupting governmental officials with money to scramble for market shares without improving on those insecure, unhealthy, and environmentally unfriendly products. It is necessary for us to add here that there is a distinction between collusion under discussion and rent-seeking in public-choice theory. Rent-seeking involves seeking to increase one's share of existing wealth instead of creating new wealth (Tullock 1967; Krueger 1974). For instance, an enterprise may enjoy preferential taxation through political lobbying. Besides, collusion is also different from "regulatory capture", a situation where an enterprise, when weighing in on policy matters, exerts influence upon the government agency responsible for regulating that particular industry so that it could get more benefits or even monopoly profits from it (Stigler 1971). Although collusion, rent-seeking, and regulatory capture are all associated with a certain form of political connection, the difference between collusion and the other two is striking: the choice of the bad mode of production in the context of collusion will undoubtedly give a telling blow to business competitors and meanwhile add luster to the government's political performance, but rent-seeking and regulatory capture may not necessarily devastate business competitors or improve the government's political performance. Of course, in many cases, the three may intersect.

Under the condition where resources are limited, which strategy will an enterprise choose? This depends on its potential earnings. On the part of the enterprise, a bad mode of production, less costly and more profitable, may eliminate the opponents; on the part of the officials, it brings them quicker and seemingly more lustrous political achievements and transfer payments such as bribes, perks, and other transfers of interests by helping the enterprise conceal information or action. On the basis of Proposition 2.1, given that the government allows collusion or that the government can profit from it, the bad mode of production is an advantageous strategy. Thus, we can make the following inference:

Proposition 6.1 *Ceteris paribus, if it is easier for an enterprise to collude with the government, the enterprise will be less motivated to make investment in research and development.*

To a large extent, the scandalous case of Rising's false incrimination against Micropoint testifies to Proposition 6.1. To maintain its market leadership of antivirus software, Rising chose to offer bribery to officials for the purpose of fabricating evidence to suppress Micropoint's innovation because it realized that it could not develop within a short period of time a new antivirus software to compete with its adversary. In the next section, systematic empirical evidence will be provided to demonstrate Proposition 6.1.

6.1.3 *Collusion and Corporate Development and Research: An Empirical Analysis*

In the past, the literature on political connections measured the degree of connection between an enterprise and a government mainly by investigating whether there are former or incumbent government officials in the corporate board or top executives. But this indicator is not applicable to the analysis of collusion for the following two reasons: former officials cannot represent the current government's will; incumbent officials can only hold a position in state-run enterprises and they represent the economic interests of the government as an investor but not the intention of the government to protect a bad mode of production. As mentioned above, collusion is a matter of agency in which information and actions are concealed, so it is hard to observe from direct evidence but can only be inferred from indirect evidence.

Collusion signifies the government's deep intervention in activities of enterprises. According to conventional wisdom, investing state capital into enterprises facilitates the government's intervention and also guarantees tunneling interests. Once the government becomes a shareholder of an enterprise, it could legally intervene in the activities of that enterprise and get economic rewards (sharing profits) and political rewards (addressing local employment problems or placing officials' relatives in the enterprise), and thus the government and the enterprise are more likely to form a union of interests. Here we take the

STA (the percentage of state-owned shares) as a proxy variable in measuring the level of collusion. According to our assumption, an enterprise with a higher percentage of state-owned shares is more likely to collude with the government and less likely to invest in R&D.

6.1.3.1 Econometric Model

Our objective is to analyze the effect of collusion on R&D, which naturally becomes the dependent variable in this model. The key explanatory variable is the STA, a proxy indicator of collusion. Besides, the major factors that influence the innovation of an enterprise also include market structure and the firm size (Schumpeter 1942; Aghion et al. 2005),[2] property right structure (Hu 2001; Nie et al. 2008), firm age (Jefferson et al. 2006), and industrial characteristics (Symeonidis 1996).[3] Hence, the econometric model is as follows:

$$
\begin{aligned}
RD_{i,t} &= \alpha + \delta STA_{i,t} + \beta_1 SIZE_{i,t} + \beta_2 MPOW + \beta_3 MCON_{i,t} \\
&\quad + \beta_4 PROP_{i,t} + \beta_5 AGE_{i,t} + \beta_6 INDU_{i,t} + \varepsilon_{i,t} \qquad (6.1)
\end{aligned}
$$

In the equation, RD stands for research and development, *STA* for the percentage of state-owned shares, *SIZE* for firm size, *MPOW* for market power, *MCON* for market concentration rate, *PROP* for property right structure, *INDU* for industrial characteristics, and *AGE* for firm age. α is a constant term, and ε is a stochastic disturbance term obeying normal distribution. In addition, we also control the dummy variable *Year*.

6.1.3.2 Data and Indicators

To test our hypothesis (Proposition 6.1), we make an econometrical analysis of the data chosen from the Chinese Industrial Enterprise Database, currently the largest enterprise-level database compiled by the National Bureau of Statistics of China, collecting basic information and financial data of 300,000 Chinese industrial enterprises. Nie et al. (2012) gave a detailed account of the present situation of its use, pointed out potential problems and then presented a correction method (see Appendix A.3). In the Database, we find some relevant information about R&D inputs in the 2001–2007 (excluding 2004) database.[4] In compliance with international practice and comparability, we choose the 2001–2007 (excluding 2004) manufacturing enterprise as the object of analysis and remove obviously unreasonable observations, including (1) those observations due to omission of sales volume, staff number, total assets, and net fixed assets; (2) those observations associated with a firm whose staff number is below 10 (because these firms lack a reliable accounting system); (3) those observations associated with the following situations where total assets are less than liquid assets, or where total assets are less than net fixed assets, or where accumulated depreciation is less than current depreciation; (4) those observations due to lack of R&D investment. In this way, we obtain 152,658 observations (which may vary slightly for different variables, see Table 6.3) from approximately 76,000 manufacturing enterprises.

We measure R&D capabilities by R&D concentration, a ratio of an enterprise's R&D investment to its sales volume; measure the level of collusion by state-owned shares, a variable which also signifies property rights (nevertheless, they will be distinguished when we regress the values); measure firm size by sales volume; measure market power by profit margin[5]; measure market concentration in 3-digit industry by HHI[6] (Herfindahl-Hirschman Index); measure property rights by the controlling shareholder, a dummy variable; measure industrial characteristics by capitalization, i.e., fixed assets per capita.

Table 6.3 A descriptive statistics of major variables of enterprises

Variables	*Indicators*	*Observations*	*Average*	*Standard deviation*	*Minimum*	*Maximum*
Innovation	R&D Concentration	152,658	0.015	0.036	1.64-e–07	0.977
Collusion	State-owned shares	152,658	0.110	0.289	0	1
Firm size	Sales volume (CNY)	152,658	261,429.5	1759,658	4.902	1.82e+08
Marker power	Profit margin	152,658	1.067	2.388	0.003	929.555
Market concentration	HHI	152,658	0.012	0.017	0.001	0.910
Age	Year	152,517	13.790	26.164	0	62
Industry characteristics	Fixed assets per capita (CNY)	152,658	117.412	236.381	0.002	17,892.62

It can be easily found from Table 6.3 which presents a descriptive statistical analysis of major variables of all the enterprises engaging in R&D activities, the R&D concentration of enterprises is far behind top MNCs in Table 6.1. The average proportion of state-owned shares is 11%, indicating that the government has a certain degree of manipulation and influence upon the manufacturing industry on the whole. Different enterprises vary greatly in size and competitions in different sectors of manufacturing also vary greatly.

6.1.3.3 Main Regression Results

According to econometric model (6.1), we construct 2001–2007 (excluding 2004) panel data of China's manufacturing enterprises, and then we regress these data by using ordinary least square (OLS) and fixed effect (FE) and we obtain the following results (see Table 6.4).

We put in model (1) only the variable STA (the proportion of state-owned shares) that reflects the level of collusion and we find the coefficient above the 10% level is significantly negative. This shows that collusion reduces R&D inputs. In order to eliminate the effect of unobservable factors, we make fixed effects regression in model (2) and control the dummy variable *Year* and still find that the coefficient of collusion above the 10% level is significantly negative. Ceteris paribus, if the proportion of state-owned shares increases by a factor of 1%, the R&D concentration will decrease by 0.1%.

Table 6.4 The effect of collusion on corporate innovation

Independent variables	*Dependent variables: Innovation*				
	OLS(1)	*FE(2)*	*FE(3)*	*FE(4)*	*FE(5)*
Collusion	−0.001*	−0.001*	−0.001*	−0.001*	0.0 002*
	(0.100)	(0.080)	(0.090)	(0.069)	(0.924)
Firm size			−3.08e−10***	−3.09e−10***	−3.37e−10***
			(0.001)	(0.001)	(0.003)
Market power			−0.00 001***	−0.0 001***	−0.0 001***
			(0.004)	(0.004)	(0.007)
Market concentration			0.0 001	0.0 001	−0.003
			(0.990)	(0.991)	(0.705)
Property rights				−0.0 005	control
				(0.216)	
Age			1.91e–06	1.94e–06	9.20e–07
			(0.665)	(0.661)	(0.856)
Industrial characeristics			control	control	control
Year		control	control	control	control
R^2	0.000	0.012	0.013	0.013	0.011
Observations	152,658	152,658	152,517	152,517	135,234

Notes The value in the parentheses () is a *p*-value; the asterisks *** and * Denote the significance level of 1 and 10% respectively

When we add other control variables to model (3), the coefficient of collusion is still significantly negative. The above regression results testify to Proposition 6.1 that collusion discourages the incentive for R&D and does harm to technological innovation. We also find that the coefficients of firm size and market power above the 1% level significantly affect the R&D concentration and that the effects of the firm size and market power are negative. This implies that, for an enterprise capable of making R&D, the larger its firm size or the stronger monopolistic capabilities, the less it will be motivated to make R&D investments. Our research negates the Schumpeterian hypothesis but testifies to Arrow's argumentation that given a specific condition, competitive enterprises could be more easily motivated to be engaged in R&D than monopolistic enterprises or monopoly could undermine innovation (Arrow 1962). Interestingly, Nie et al. (2008) ran a random effect Tobit model and put all manufacturing enterprises into the research sample and found that there was an inverted **U** relationship between market structure and innovation, i.e., a certain degree of market power prompts incentives for R&D but a much more degree of market power undermines incentives for R&D. Market concentration is not significant perhaps because market power has captured the effect of market structure on R&D. *Age* is not significant either, indicating that there is no substantial difference between new and old enterprises in R&D investment.

In model (4), we control property rights and put in the proportion of private shares and find the results remain unchanged. This shows that privatization itself will not be able to alter collusion and governmental intervention in enterprises. In model (5), we eliminate all the samples of state enterprises and find that the variable collusion is no longer significant. This implies that crowding-out effect of R&D caused by collusion is more obvious in state-run enterprises and less obvious in nonstate-run enterprises. Our research implies that, in the process of privatizing state-run enterprises, much emphasis should be laid on those aspects that are more likely to result in collusion so as to restrain collusion and weaken governmental intervention in the production activities of enterprises.[7]

6.2 Collusion Distorting Total Factor Productivity

6.2.1 *Collusion and Productivity: A Theoretical Analysis*

"Productivity isn't everything, but in the long run, it is almost everything," says Paul Krugman (1997), the 2008 Nobel Memorial Prize laureate in Economic Sciences to highlight the real importance of productivity. Therefore, our research on the impact of collusion on enterprises is focused on the impact of collusion on productivity. Technological innovation is not the sole factor that determines productivity, other factors such as management, market environment, and institutional environment also affect productivity.

Against the background of the decentralization, those enterprises colluding with local governments can take the advantage of preferential taxes, subsidies, and cheaper productive factors (such as credit funds) granted by local governments, but they also pay a painful price in the following aspects: (i) helping local governments address unemployment because a low rate of unemployment is always an indicator or rather an assessment criterion of excellent political performance; (ii) recruiting some governmental officials' family members or relatives, who are actually freeloaders loafing on the job or employed through political *guanxi*; (iii) offering officials perks and other forms of bribery, such as providing them with overseas trips or arranging overseas schooling for their children; (iv) bearing the cost of the time squandered on establishing *guanxi* with officials; (v) bearing the cost spent on evading the supervision from the central government or the possible mass media disclosure.[8] What's worse, compared with a better mode of production, a worse mode of production signifies that limited resources (including human resources) are put into nonproduction activities such as political connections, corruption, and rent-seeking rather than production activities such as research and development, improving management efficiency, and expanding market. Therefore, from the perspective of Pareto Optimality or Efficiency, collusion between the local government and enterprise will result in deadweight loss and thus reduce productivity. This logic reasoning is similar to the logic that corruption reduces productivity (Murphy et al. 1991; Djankov et al. 2002). But as mentioned above, collusion is not necessarily concomitant with corruption, although it is very likely to be so.

Recent economics papers term this phenomenon as "misallocation of resources", i.e., limited resources are not allocated either to best efficient enterprises or to best efficient production activities. In empirical observations, misallocation of resources is often reflected as a big difference in productivity and return on invested capital (ROIC) between different enterprises (Banerjee and Duflo 2005). It is believed among economics experts that misallocation of resources has a more significant effect on wealth per capita or productive efficiency than on technological innovation. Alfaro et al. (2008) used approximately 24 million data collected from 79 countries (including China) and found that misallocation of resources between enterprises fell out nicely from the difference of per capita income. Hsieh and Klenow (2009), by combining a monopoly-competition model with data of enterprises, made a sound calculation that, if China and India allocate resources to high-productivity as the USA does, the TFP of the two countries will be increased by 30–50% and 40–60% respectively. Nie and Jia (2011), by using different methods, calculated the degree of misallocation of resources of Chinese manufacturing enterprises. If seen from another perspective, collusion actually results in misallocation of resources in different enterprises and in different activities of the same enterprise. If the local governments at different levels allocate resources to enterprises adopting a good mode of production, or each enterprise allocates more of its resources to production activities instead to rent-seeking, the efficiency will be greatly improved. Thus, we get the following proposition.

Proposition 6.2 *Collusion between local governments and enterprises probably results in misallocation of resources to nonproduction activities, thus reducing productivity of enterprises.*

6.2.2 *Collusion and Productivity: Evidence*

Estimating TFP

In this section, we use the 1999–2007 Chinese Industrial Enterprise Database to test Proposition 6.2. Since the objective is to test the effect of collusion on the productivity of enterprises, productivity is set as the dependent variable. The major indicator to measure productivity is TFP. In the literature on contemporary theory of international industrial organization, it is a customary practice to use OP method to estimate TFP (Olley and Pakes 1996),[9] because only OP method can solve the problems of simultaneity and sampling at the same time in the estimation of TFP. In Appendix A.4, I compare the methods that are employed to estimate TFP and briefly explain how OP method works. Readers can also refer to Nie and Jia (2011) for a comprehensive understanding of the OP method.

According to the customary practice, we use sales volume to measure output,[10] the number of staff to measure workforce, reported value to denote intermediate inputs, and net fixed assets to measure capital. Perpetual inventory method is used to calculate investment, i.e., $I_{it} = K_{it} - (1 - \delta)K_{it-1}$, among which depreciation rate δ is set at 15%.[11] In order to get a TFP in its economic sense instead of in its accounting sense, we should also use price deflator series. Considering the cross-regional differences of price fluctuations, sales volume and intermediate inputs are deflated on the basis of PPI (Producer's Price Index for Manufactured Products) of each province, and capitals are deflated on the basis of price index of investment in fixed assets of each province.[12] Price indices are collected from *China Statistical Yearbook* of all years. We use the consolidated data and estimate TFP by means of OLS and OP respectively. Combining econometrical theory and comparing the estimated slopes of OLS and OP, we find estimated results based on OP method are more agreeable with theoretical expectations.

Some modifications should also be done with the dependent variable TFP. Due to great differences across different sectors within the manufacturing industry itself, the TFP, if estimated on the basis of all manufacturing samples, cannot truly reflect these cross-sector differences. Therefore, we follow the practice by Hsieh and Klenow (2009) and divide the TFP of each enterprise into the average TFP of the whole sector and take the logarithm of it, i.e., $RTFP_{it} = \log(TFP_{it}/TFP_t)$. This implies that for those enterprises whose TFP is above the average value, the standardized TFP is positive; otherwise it is negative. This solution removes the effect of cross-sector differences of TFP on an individual enterprise's TFP.

6.2.2.1 Econometric Model

The key explanatory variable is collusion. Besides the proportion of state-owned shares, we include another proxy indicator of collusion, i.e., government subsidy. We assume that an enterprise more inclined to collude with a local government is more likely to be subsidized; therefore, we use SUB (the proportion of subsidy to sales volume) to denote the degree of subsidy. It must be expressly stated that private enterprises may sometimes get subsidy from the government although most of it is granted to state-owned enterprises. For instance, some private enterprises can get government funds for innovative projects or subsidized loans.

Besides the key explanatory variable collusion, other factors also affect TFP. On the basis of the research by Keller and Yeaple (2009) and Syverson (2010), we survey the data structure and take into consideration the following factors.

1. **R&D** Technological advancement increases productivity and its major source is R&D investment, which in this case is expressed as RD (denoting the proportion of R&D investment to sales volume, i.e., R&D concentration). The R&D data in 2004 is omitted.
2. **Market structure** According to Schumpeterean hypothesis (Schumpeter 1942), monopolistic enterprises have stronger innovative capabilities or be more efficient in introducing advanced technology to improve productivity. However, on the other hand, theoretical researches show that competition prompts enterprises to make innovation. The two arguments above do not conflict with each other because the former involves only a single market power but the latter involves the market environment of the whole sector. We use the enterprise's profit margin to express **MPOW** (denoting

market power), use HHI to express MCON (denoting market concentration) in 4-digit industry. In order to avoid the reverse effect of productivity on market structure, we follow Keller and Yeaple (2009) and take first order lag of profit margin and HHI.[13]

3. **Export** The international trade theory based on firm heterogeneity argues that, due to selection effect or export-learning effect, export enterprises tend to have a higher productivity (Melitz 2003). We add a dummy variable EXPO (denoting export) to express export (=1) or nonexport (=0).
4. **FDI spillover** FDI will theoretically cause a positive spillover effect on domestic enterprises because foreign investment enterprises are often regarded as being highly efficient in developing countries such as China. We use the proportion of sales volume of the FDI enterprises in 4-digit industry to the aggregate sales volume of the whole sector to express the degree of FDI.
5. In addition, in view of the fact that the proportion of labor to capital has an effect on an enterprise's absorption of technology and its TFP, we take the logarithm of per capita fixed assets (denoted by *K_L*.) to express the degree of capitalization. This indicator also signifies barriers to entry or competitive insulation and therefore it controls industry factors to a certain degree. Year. is also set as a dummy variable. Therefore, our econometric equation is constructed as follows:

$$
\begin{aligned}
TFP_{i,t} = \alpha &+ \delta_1 SAT_{i,t} + \delta_2 SUB_{i,t} + \beta_1 RD_{i,t} \\
&+ \beta_2 MPOW_{i,t} + \beta_3 MCON_{i,t} + \beta_4 EXPO_{i,t} \\
&+ \beta_5 FDI_{i,t} + \beta_6 K\_L_{i,t} + \varepsilon_{i,t}
\end{aligned}
$$

6.2.2.2 Major results

Considering the data structure is panel data, we regress these variables using fixed effects model. The results are as follows (Table 6.5).

The major regression equation is model (1) because the dependent variable is OP-estimated $RTFP^{OP}$. The two explanatory variables *STA* and *SUB* at the level of 1% are significantly negative. This signifies that collusion significantly reduces TFP, and this is accordant with our theoretical expectation (Proposition 6.2). As mentioned above, the reform of privatization of state-run enterprises is still necessary in those fields where collusion is very likely to happen. In contemporary literature, government subsidy is often regarded as one of the major sources of misallocation of resources (Restuccia and Rogerson 2008) because it may probably result in the problem of soft budget constraint

Table 6.5 The effects of collusion on total factor productivity

Independent Variables	*Dependent variable*			
	(1)$RTFP^{OP}$	*(2)TFP^{OP}*	*(3)$RTFP^{OLS}$*	*(1)$GTFP^{OP}$*
STA.	−0.009*** (0.000)	−0.018*** (0.000)	−0.019*** (0.000)	−0.008*** (0.000)
SUB.	−0.031*** (0.000)	−0.048*** (0.000)	−0.069*** (0.000)	−0.039*** (0.000)
R&D	−0.334*** (0.000)	−0.593*** (0.000)	−0.765*** (0.000)	−0.257*** (0.000)
MPOW. -1	0.0→003** (0.031)	0.001** (0.020)	0.001** (0.014)	−0.0 004* (0.051)
MCON $_{.-1}$	−0.018* (0.079)	1.26 (0.209)	0.026 (0.325)	0.011 (0.541)
FDI.	−0.031*** (0.000)	−0.005 (0.523)	0.0 004 (0.997)	0.0 006 (0.935)
EXPO	−0.001 (0.598)	−0.004* (0.063)	0.002 (0.403)	−0.0 003 (0.877)
K_L	control	control	control	
	control	control	control	
R^2	0.007	0.023	0.026	0.002
Observations	1008,100	1008,100	1008,100	1007,457

Notes The value in the parentheses () is a *p*-value; the asterisks *** and * Denote the significance level of 1 and 10% respectively

(Kornai et al. 2003).[14] Our finding testifies to this from the perspective of collusion. In addition, this shows that hardening the budget constraint of an enterprise can reduce the likelihoods of collusion and thus improve its total factor productivity.

In model (1), R&D concentration (*RD*) is in significantly negative correlation with TFP. This seems to be contradictory with our theoretical prediction, but it is really accordant with China's current situation because R&D investment of Chinese manufacturing enterprises is not totally motivated by market but by tax evasion to such a large extent that the R&D investment of private enterprises increases by a wide margin due to the issuance and implementation of preferential tax measures (Nie et al. 2008).[15] As a matter of fact, when we regress R&D concentration with profit margin of enterprises, the two are in significantly negative correlation. This also tests our assumption from another perspective. The signs (plus or minus) of *MPOW* and *MCON* are also accordant with our theoretical expectation, i.e., enterprises with a stronger market power could earn higher profit margin and are more capable of innovating or introducing advanced technology to improve TFP; stiffer market competition implies that enterprises faces stronger external constraints and are more inclined to be motivated to improve TFP. The FDI spillover effects are minus because FDI enterprises help domestic enterprises to improve TFP through the diffusion of technology on the one hand but crowd out market share of domestic enterprises through competition on the other. When competition effects surpass technological diffusion effects, the FDI spillover effects are minus (Zhang 2005). Interestingly, we find that *EXPO* is not in significantly negative correlation with TFP. Is this in variance with the prevailing international theory based on firm heterogeneity? An intuitive interpretation is: many Chinese export enterprises actually process raw materials on clients' demands, assemble parts for the clients and process according to the clients' samples or engage in compensation trade, or processing trade, so their productivity remains low. As a matter of fact, Dai et al. (2011) made a research by matching the Chinese Industrial Enterprise Database with Customs House Database and found that "productivity paradox" does not exist after they eliminated the sample of processing trade enterprises. In addition, export enterprises can enjoy the advantage of export tax rebates or drawbacks, which further distorts the relationship between export and productivity.

In order to test the robustness of the results above, we replace the dependent variables in models (2), (3) and (4) with OP-estimated TFP^{OP}, OLS-estimated TFP^{OLS} and OP-estimated $GTFP^{OP}$ respectively. The results show that *STA*, *SUB*, and *RD* at the level of 1% are significantly negative. The *MPOW* effects almost remain unchanged, but *MCON*, FDI spillover effects and export effects are basically insignificant. To conclude, *STA* and *SUB*, the two major indicators of collusion robustly constitute the major reasons for the decline of TFP. This shows that our results are extremely robust.

6.3 Collusion and Enterprises' FDI

6.3.1 An Overview China's FDI Trend

At present, China is not only one of the largest FDI (foreign direct investment) recipients but also one of the major ODI (outbound direct investment) exporters. According to the statistics of the United Nations Conference on Trade and Development, the FDI inflow that China attracted in 2010 amounted to $101 billion, ranking second in the world; the outflow of ODI amounted to \$68.81 billion, ranking fifth in the world (UNCTAD 2011). Fig. 6.1 makes a comparison of FDI inflows of major countries or regions.

According to *Statistical Bulletin of China's Foreign Direct Investment in 2010* released by Ministry of Commerce of the PRC, the outflow of China's ODI has the following characteristics.

Firstly, the outflow of ODI increases fast and has successively hit record highs in recent years. In 2010, China's outflow of ODI surpassed Japan and the UK for the first time. Between 2002 and 2010, its annual growth rate was as high as 49.9%.

Secondly, China's ODI is mainly injected into overseas banking, manufacturing, and mining. In 2010, the outflow of ODI through merger and acquisition amounted to \$29.7 billion, accounting for 43.2% of the total. ODI is mainly invested in banking (12.5%), wholesaling and retailing (9.8%), mining (8.3%), transportation, warehousing and postal services (8.2%), manufacturing (6.8%), construction (2.4%), and real estate (2.3%).

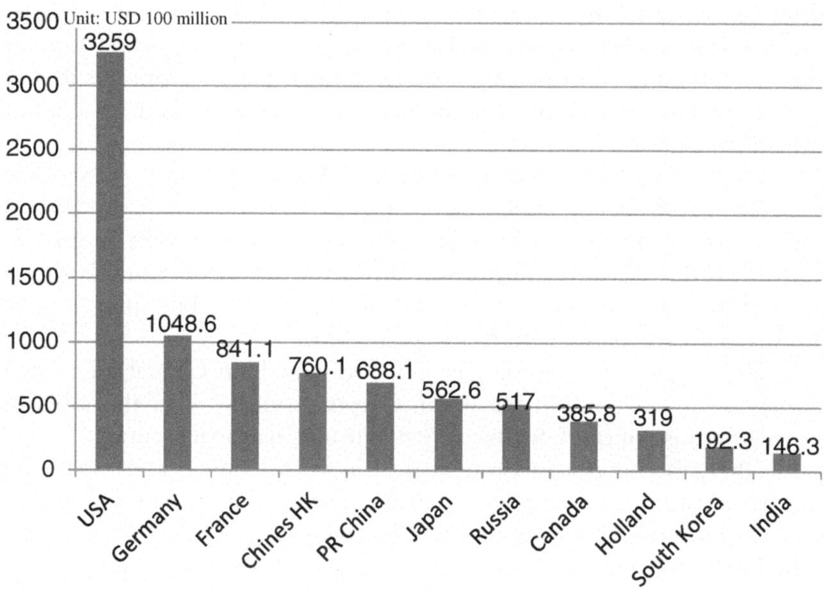

Fig. 6.1 FDI inflows of major countries or regions in 2010

Thirdly, from the regional distribution, China's ODI is relatively concentrated. In 2010, China's ODI was concentrated in its top nine destinations, each of which received at least $1 billion. The sum of the investment in these nine destinations accounted for 84% of its aggregate ODI all over the world. The top nine destinations were Hong Kong (56%), British Virgin Islands (8.9%), Cayman Islands (5.1%), Luxembourg (4.7%), Australia (2.5%), Sweden (1.9%), the US (1.9%), Canada (1.7%), and Singapore (1.6%).

Fourthly, central enterprises dominate the ODI market. In 2010, the nonfinancial outbound direct investment of central enterprises amounted to $42.44 billion, accounting for 70.5% of the total. In terms of investment volume, the top 20 enterprises were all central enterprises, including Sinopec, PetroChina, China National Offshore Oil Corporation (CNOOC), China Resources (Holdings) Co., Ltd., China COSCO Shipping Corporation Limited, China Oil & Foodstuffs Corporation, and Aluminum of China Corporation Limited.

6.3.2 *Obstruction of Overseas Acquisition and Merger*

Although some Chinese enterprises have succeeded in overseas merger and acquisition (M&A) with a fanfare of publicity, the proportion of failures of bidding remains high. Some established enterprises suffered repeated frustrations in overseas merger and acquisition. Here is a list of cases of overseas FDI failures.[16]

In June 2005, China National Offshore Oil Corporation (CNOOC) made an $18.5 billion cash offer for American oil company Unocal Corporation, topping an earlier bid by Chevron Texaco. On August 2, 2005, CNOOC, facing the pressure from the US government's direct intervention, announced that it had withdrawn its bid after several months of seesaw battle with Chevron Texaco.

In 2007, Huawei Technologies Co. Ltd. and Bain Capital attempted to make a $2.2 billion bid for 3Com Corporation, but later this attempt aborted because of the US government's fear of national security.

In 2009, the biggest investment deal in both Chinese and Australian corporate histories, a $19.5 billion deal between Aluminum Corporation of China Limited (Chinalco) and Rio Tinto Group (Australia) collapsed.

In 2010, Tencent's bid for ICQ, an instant messaging computer program, aborted.

In 2010, China National Offshore Oil Corporation (CNOOC) and Ghana National Petroleum Corp jointly made a $5 billion bid for Ghana Jubilee Oilfield but failed.

In 2010, SinoChem Group's and Singaporean Temasek Holdings jointly attempted to make a $50 billion bid for Canadian potash but failed.

In 2010, Huawei made a bid for Motorola, but its adversary Nokia Siemens Networks won it at a lower price than Huawei's $1.2 billion.

In 2010, General Motors (GM) declared at its headquarters in Detroit that Sichuan Tengzhong Heavy Industries Co. Ltd. failed to take over it as schedule and the deal aborted.

In January 2011, Bright (Food) Group Co., Ltd. withdrew from its negotiation with GNC (General Nutrition Corporation) Holdings Inc., because both parties failed to come to an agreement on price and other terms.

In March 2011, Bright (Food) Group Co., Ltd. verified that Yoplait, the second largest French yoghurt maker of the world had entered into exclusive negotiations to buy a majority stake, indicating that its earlier highest bid for a 50% stake failed.

In February 2011, Huawei was barred from US contracts and acquisitions. A US congressional investigation panel demanded, on security grounds, that Huawei should strip the technology assets acquired from 3LeafSystem. Later, Huawei announced it withdraw from its attempt to take over 3LeafSystem, marking its second failure to tap the North American market.

Why did the above cross-country takeover fail? Careful readers may find an important characteristic: the Chinese enterprises that initiated merger and acquisition are all large central enterprises or large enterprises closely associated with the central government. Is there a definite relation between this characteristic with repeated failures of overseas merger and acquisition?

6.3.3 Double-Edged Sword Effect of Collusion

From Statistical Bulletin of China's Foreign Direct Investment by MOFCOM and cases of failure in overseas takeover, we find that the initiators of overseas merger and acquisition are central enterprises or enterprises closely associated with the government. This impels us to think it over from the perspective of collusion.

Imagine a domestic enterprise which has established a close relation with a local government through nationalization or through other forms of interest conveyance. It may collude with the local government to choose a bad mode of production such as administrative monopoly, subsidy, and suppressing competitors to seek illegitimate gains.[18] In this way, this enterprise can quickly develop and occupy the domestic market before it enters the overseas market. Undoubtedly, at least in the short run, collusion can help a Chinese market to enter a foreign market. As a matter of fact, it is very important for an enterprise to gain the government's support if it plans to expand its market in a country with a different institutional environment and cultural background. This explains why the enterprises that are able to make FDI abroad are central enterprises: because only central enterprises have more and better opportunities to collude with the government than private cousins.

However, collusion has a double-edged sword effect. The government's support is an advantage to the domestic enterprise but a disadvantage to the host country's government and enterprises which therefore will try every conceivable means to obstruct the merger and acquisition by those Chinese enterprises in collusion with the government. To the host country, subsidy means an unfair competition, but "national security" is just a political camouflage.

A recent case associated with Huawei verified our reasoning.

On October 8, 2012, the US Congress released Investigative Report on the US National Security Issues Posed by Chinese Telecommunications Companies Huawei and ZTE, asserting that the two Chinese giant telecom firms had provided the Chinese intelligence agencies with the opportunities to intervene in the US telecommunication network. The report also claimed that the two firms neither cooperated with the US congressional 1-year-long investigation nor explained their commercial interests in the US and their relations with the Chinese government. The leader of the US congressional investigation panel warned that the US House Intelligence Committee should keep alert on Huawei and ZTE's commercial expansion in North America and that the US private enterprises should be continuously informed of the potential espionage threats from China.

On September 13, 2012, the top executives of Huawei and ZTE were questioned by US congressmen at the three-hour-long hearing, suspected of thieving American corporation's secrets and posing security risk to American network facilities. The top executives were even demanded to explain why such Chinese private enterprises as Huawei and ZTE set up a CPC committee.

In the above case, ZTE (Zhongxing New Telecommunications Equipment Co., Ltd), China's largest telecommunications equipment and systems listed company initially incorporated by a group of investors closely associated with the Ministry of Aerospace, creates a new business model as "state-owned and private-operating" economic entity. Its aerospace-backed background was suspected of benefiting from collusion with the government. Huawei, the second largest telecommunications equipment supplier of the world, was inextricably tied up with the government although it is a private enterprise. According to the information released to the public, it has established more than 100 joint ventures with each of the telecommunications management bureaus in 18 provinces since 1996. These telecommunications management bureaus, with a double identity of Huawei's clients and shareholders, formed a coalition of interests with Huawei. Although its expansion strategy was controversial in industrial circles, it quickly occupied the telecom markets of many provinces with the help of provincial governments, solved the problems of payment collection and shortage of funds (Cheng and Liu 2004). Furthermore, Huawei's secret corporate governance structure and the grapevines about top executives' political backgrounds cannot remove the suspicion that it benefited from collusion. Undoubtedly, be it Huawei or ZTE, collusion is a great disadvantage for them to expand the overseas market.

6.4 Summary

This chapter discusses the effects of collusion on corporate behaviors and performances. We have the following findings:

Firstly, collusion obstructs the innovation of enterprises. Since an enterprise can obtain monopoly profit and defeat its competitors simply through the means of collusion, it has less motivation to be engaged in R&D. Basing our research on the Chinese Industrial Enterprise Database, we use the proportion of state-owned shares in enterprises to measure the level of collusion and the capital input of R&D to measure technological innovation and discover that collusion significantly reduces motivation for technological innovation.

Secondly, collusion distorts enterprises' TFP. Collusion compels an enterprise to allocate its limited resources to nonproduction activities such as rent-seeking, bribery, and bureaucracy, thus resulting in misallocation of resources and lowering enterprises' efficiency. Collusion measured on the basis of state-owned shares and subsidy significantly lower enterprises' TFP.

Thirdly, collusion is a disadvantage for Chinese enterprises to participate in international competition. Collusion means subsidy or some other special services provided by the government. This is often regarded as an unfair competition by foreign governments and firms.

The policy implication of this chapter consists in impartial treatment to all enterprises. Granted, collusion can promote a quicker economic growth in the short run, but that is extensive, unsustainable growth without technological innovation. Therefore, the government should do justice to all enterprises, encouraging them to put more investment in R&D. To be specific, the support policies are as follows: (1) subsidize technological innovation, grant innovative enterprises with subsidized loans, and particularly protect intellectual property rights; (2) reduce the proportion of state-owned enterprises and the proportion of state-owned shares in such sectors which are likely to be associated with collusion as environment pollution, coal mining, and telecommunications equipment and systems, and meanwhile promote the process of privatization; (3) differentiate the normal services from abnormal collusion, encourage enterprises to develop on its own, and support private enterprises to expand overseas markets.

Notes

1. Total factor productivity, often recognized as a variable representing the amount of output of factors (in most cases, science and technology) except labor and capital, is a commonly used index to measure the efficiency of an enterprise (Syverson 2010).
2. It is assumed, according to the Schumpeterian hypothesis, that bigger enterprises are more capable of making innovation than smaller ones because the former enjoy relative advantages in economies of scale, risk sharing and financing channels; and that monopolistic enterprises have stronger innovative capabilities because R&D needs continuous financial support.
3. For literature on corporate innovation and R&D, refer to Symeonidis (1996), Subodh (2002) and Nie et al. (2008).

4. The year 2004 was an economic census year and no investigation was made into enterprises' R&D inputs and export.
5. The computational method of profit margin is as follows: profit margin = sales volume/(sales volume—operating profit). The more the profit margin, the stronger the market power, or the more monopoly profit it can get.
6. The computational method of HHI is as follows:HHI = $\sum (s_i/S)^2$, i.e. the squared value of the ratio of an enterprise's market share to the whole particular industry. The lower the value of *HHI*, the more fierce the market competition.
7. However, this does not imply that there is no possibility of local governments' collusion with private enterprises. Theoretically, so long as there is exchange of interests, an alliance of collusion may be formed. For instance, in the Dalian PX project (see Chap. 2), the holding party the Fujia Dahua Co. Ltd. undertaking the PX project was Fujia Investment Corporation, a private enterprise holding 51% of the shares.
8. For instance, on September 20, 2008, a coal worker was killed in an accident due to negligence of work safety in a coalmine of Hobaoganhe Coal Co. Ltd., Hongdong County, Shanxi province, but the coal company concealed the fact. By the evening of September 25, correspondents who had already known the score got a large sum of hush money varying between thousands and tens of thousands of Chinese yuan at an office of the company. This accident was later exposed to the public by a media from another province. For full report of this accident, see "Coalmine accident and hush money exposed by peers", link: http://www.caijing.com.cn/2008-11-27/110032378.html.
9. As a matter of fact, in Syverson (2010), OP method is the only important method to estimate productivity.
10. EVA (economic value added)-based TFP is not estimated because there is no variable EVA for nearly half of the enterprises.
11. Some literature sets depreciation rate as 10% or other percentages. Using different depreciation rates and price deflation methods will not change the basic results.
12. The National Bureau of Statistics of China (NBSC) provided the Producer's Price Index for Manufactured Products of 14 sectors of the manufacturing industry, but our sampling of manufacturing industry include 30 2-digit sectors. The two are not in one-to-one correspondence. In addition, the NBSC does not provide Price Indices of Investment in Fixed Assets based on different sectors.
13. When we take first order lag of all explanatory variables, the results almost remain unchanged.

14. The concept of soft budget constraints means that the subsidies and capitals that enterprises receive from government are not rigid. The less efficient an enterprise is, the more subsidies it could receive from government. In contract theory, this is an issue of dynamic moral hazard.
15. In 1999, the State Administration of Taxation issued Administrative Measures of Pre-Tax Deduction for Enterprise R&D Expenses, stipulating that any state-owned or collective enterprise which increases its R&D expenses by 10% than the previous year can use 50% of its R&D expenses to deduct its corporate income tax. At the end of 2003, this preferential treatment trickled down to all industrial enterprises with sound financial system.
16. The original source of each piece of the following information is not clear. Nevertheless, the cases are widely spread throughout all kinds of international mass media.
17. When it is assumed that an enterprise chooses a certain bad mode of production, it does not necessarily mean that this enterprise will not take a good mode of production.
18. When it is assumed that an enterprise chooses a certain bad mode of production, it does not necessarily mean that this enterprise will not take a good mode of production.

References

Aghion, Philippe, Nicolas Bloom, Richard Blundell, Rachel Griffith and Peter Howitt. "Competition and Innovation: An Inverted U Relationship". *Quarterly Journal of Economics*. 2005, 20 (2): 701–728.

Alfaro, Laura. Andrew Charlton: Fabio Kanczuk. "Plant-Size Distribution and Cross-Country Income Differences". NBER Working Paper No. 14 060, 2008.

Arrow, Kenneth. "Economic Welfare and the Allocation on Resources for Invention". In Nelson, R. R. eds. *The Rate and Direction of Inventive Activity*, Princeton: NBER, 1962.

Banerjee, Abhijit V., Esther Duflo. "Growth Theory through the Lens of Development Economics". In Philippe Aghion and Steven Durlauf. eds. *Handbook of Economic Growth*, edition 1, 1 (1). Amsterdam: North-Holland, 2005.

Caselli, Francesco. "Accounting for Cross-Country Income Differences", in Philippe Aghion and Steven Durlauf eds. *Handbook of Economic Growth*. Amsterdam: Elsevier, 2005, volume 1, pp. 679–741.

Cheng, Dongsheng and Lili Liu. *A True Story of Huawei*. Beijing: Contemporary China Publishing House. 2004. 程东升, 刘丽丽. 华为真相. 北京: 当代中国出版社. 2004.

Chen, Xinyuan, et al. "Judiciary Independence and the Enforcement of Investor Protection Laws: Market Responses to the '1/15' Notice of the Supreme People's Court". *China Economic Quarterly*. 2010 (1): 1–28.

Claessens, Stijin, Eric Feijen and Luc Laeven. "Political Connections and Preferential Access to Finance: The Role of Campaign Contributions". *Journal of Financial Economics*. 2008, 88: 554–580.

Dai, Mi, Maitra, Madhura and Yu, Miaojie. "Unexceptional Exporter Performance in China: The Role of Processing Trade". Peking University working paper. 2011.

Djankov, Simeon, Rafael La Porta, Florencio Lopez-DeSilanes, and Andrei Shleifer. "The Regulation of Entry". Quarterly Journal of Economics, 2002, 117 (1): 1–37.

Faccio, Mara. "Politically Connected Firms". *American Economic Review*. 2006, 96 (1): 369–386.

Fan, Joseph, T. J. Wong and Tianyu Zhang. "Politically connected CEOs, corporate governances, and post-IPO performance of China's Newly Partially Privatized Firms". *Journal of Financial Economics*. 2007, 2: 330–357.

Fisman, Raymond. "Estimating the Value of Political Connections". *American Economic Review*. 2001, 91: 1095–1102.

Hall, Robert E. and Charles I. Jones. "Why Do Some Countries Produce So Much Output per Worker than Others?" *Quarterly Journal of Economics*. 1999, 114: 83–116.

Hsieh, Chang-Tai and Peter J. Klenow. "Misallocation and Manufacturing TFP in China and India". *Quarterly Journal of Economics*, 2009, 124 (4): 1 403–1 448.

Hu, Albert G. Z. "Ownership, Government R&D, Private R&D, and Productivity in Chinese Industry". *Journal of Comparative Economics*. 2001, 29 (1): 136–157.

Jefferson, G.H., Huamao Bai, Xiaojing Guan, and Xiaoyun Yu. "R and D Performance in Chinese Industry". *Economics of Innovation and New Technology*. 2006, 15 (4/5): 345–366.

Keller, Wolfgang and Stephen R. Yeaple. "Multinational Enterprises, International Trade, and Productivity Growth: Firm-Level Evidence from the United States". *Review of Economics and Statistics*. 2009, 91 (4): 821–831.

Kornai, Janos, Eric Maskin and Gerard Roland. "Understanding the Soft Budget Constraint". *Journal of Economic Literature*. 2003, 41 (4): 1 095–1 136.

Krueger, Anne. "The Political Economy of the Rent-Seeking Society". *American Economic Review*. 1974, 64 (3): 291–303.

Krugman, Paul. *The Age of Diminished Expectations: U.S. Economic Policy in the 1990s*. 3rd edition. Cambridge: MIT press, 1997.

Luo, Danglun and Qingquan Tang. "The Performance of Institutional Environment Evidence from China's Private Listed Companies". *Economic*

Research Journal. 2009, (2): 106–118. 罗党论, 唐清泉. 市场环境、中国民营上市公司制度环境与绩效问题研究. 经济研究. 2009 (2): 106–118.

Melitz, Marc. "The Impact of Trade on Intra-industry Reallocations and Aggregate Industry Productivity". Econometrica. 2003, 71 (6): 1 695–1 725.

Murphy, Kevin, Andrei, and Robert Vishny. "The Allocation of Talent: Implications for Growth". *Quarterly Journal of Economics*. 1991, 106 (2): 503–530.

Nie, Huihua, Ting Jiang, and Rudai Yang. A Review and Reflection on the Use and Abuse of Chinese Industrial Enterprises Database. The Journal of World Economy. 2012(5): 142–158. 聂辉华, 江艇和杨汝岱,《中国工业企业数据库的使用现状和潜在问题》,《世界经济》, 2012(5): 142–158.

Nie, Huihua and Ruixue Jia. "China's Manufacturing TFP and Misallocation of Resources". *The Journal of World Economy*. 2011, (7): 27–42. 聂辉华, 贾瑞雪. 中国制造业企业生产率与资源误置. 世界经济, 2011, 34 (7): 27–42.

Nie, Huihua, Songtao Tan and Yufeng Wang. "Innovation, Firm Size and Market Competition: An Analysis Based on Panel Data of Chinese Enterprises". *The Journal of World Economy*. 2008, 31 (7): 57–66. 聂辉华, 谭松涛, 王宇锋. 创新、企业规模和市场竞争——基于中国企业层面面板数据的分析. 世界经济, 2008, 31 (7): 57-66.

Olley, G. Stevan and A. Pakes. "The Dynamics of Productivity in the Telecommunications Equipment Industry". *Econometrica*, 1996, 64 (6): 1 263–1 297.

Restuccia, Diego, Richard Rogerson. "Policy Distortions and Aggregate Productivity with Heterogeneous Establishments". *Review of Economic Dynamics*. 2008, 11 (4): 707–720.

Schumpeter, Joseph. *Capitalism, Socialism, and Democracy*. New York: Harper & Brother, 1942.

Stigler, George. "The Theory of Economic Regulation". *Bell Journal of Economics and Management Science*. 1971, 3: 3–18.

Subodh, Kandamuthan. "Market Concentration, Firm Size, Innovative Activity: A Firm-level EconomSic Analysis of Selected Indian Industries under Economic Liberalization". WIDER, Discussion Paper, No. 108, 2002.

Symeonidis, George. "Innovation, Firm Size and Market Structure: Shumpeterian Hypotheses and Some New Themes". OECD working paper no. 32643, 1996.

Syverson, Chad. "What Determines Productivity?" NBER working paper No. 15712, 2010.

Tullock, Gordon. "The Welfare of Costs of Tariffs, Monopolies and Theft". *Western Economic Journal*. 1967, 5 (3): 224–232.

UNCTAD. "World Investment Report 2010". 2011, http://unctad.org/en/pages/PublicationArchive.aspx?publicationid=753 (Re-accessed by the translator October 6, 2016).

Wu, Wenfeng, Wu Chongfeng and Liu Xiaowei. "The Governmental Backgrounds of Top Executives in Chinese Private Listed Companies and the Value of Companies." *The Research of Economics*. 2008 (7): 130–141.

Yang, Qijing. "The Growth of Enterprises: To Build Political Connections or Capability?" *Economical Research Journal*. 2011, (10): 54–66. 杨其静. 企业成长: 政治关联还是能力建设? 经济研究, 2011 (10): 54-66.

Yu, Ming Gui and Hongbo Pan. "Political Connections, Institutional Environment and Bank Loans of Private Enterprises". *Management World*. 2008, (8): 9–21. 余明桂, 潘红波. 政治关系、制度环境与民营企业银行贷款. 管理世界, 2008, (8): 9-21.

Zhang, Haiyang. "R&D Duality, FDI Activities, and the Productivity Growth of Chinese Enterprises". *Economic Research Journal*. 2005, (6): 107–117. 张海洋. R&D两面性、外资活动与中国工业生产率增长. 经济研究. 2005, (6): 107–117.

CHAPTER 7

The Impact of Collusion on Government Behaviors

This chapter is mainly concerned about the impact of collusion on government behaviors. Our theoretical analysis shows that collusion exacerbates corruption and soft budget constraints, and then we provide a case study and make regression results to demonstrate our theory. Section 7.1 discusses the relation between collusion and corruption; Sect. 7.2 analyzes the impact of collusion on soft budget constraints; Sect. 7.3 is a summary.

7.1 The Relationship Between Collusion and Corruption

7.1.1 Collusion and Corruption: A Theoretical Analysis

Collusion, in essence, is a form of cooperation between a local government and an enterprise to accelerate economic growth and gloss political performances under the double conditions of information asymmetry and GDP-oriented assessment criteria. Collusion is not always concomitant with corruption, and the "Tieben Event"[1] is a case in point, which is purely caused by a local government's impulse to add luster to its political performances. Even so, the enterprise participating in this collusion has to evade the inspection of a higher local government (an enterprise at city level or below may be inspected by a provincial government agency, for instance) or the central government, avoid the possible

H. Nie, *Collusion, Local Governments and Development in China*,
DOI 10.1007/978-981-10-5059-6_7

disclosure by the mass media and pacify or mollify local residents' resentment against its illicit requisition of land and emission of pollutants. It is almost impossible for an enterprise choosing a bad mode of production to cope with the constant inspection and pressure of public opinions without seeking umbrella of protection from different levels of local governments, which in this case may naturally be corrupted. Let us imagine the other way around, when an official has no or much less gray income than ever before, he would attempt to improve his political performances so as to get more marginal earnings, which, compared with gray income, seem almost negligible. Nevertheless, under the system of centralization, political performances are often attributed to the "correct and wise" leadership of the head of a government agency, whichever level it is. This implies that any of those officials who populate a local government[2] are likely to get involved in corruption. In the realm of economics research, corruption is often defined as abuse of public power or authority by the government officials for personal interests (Svensson 2005).

Take for instance the coalmine accidents across the country. On March 14, 2005, 18 coal workers were killed in a gas explosion in Fuchun Coalmine in Heilongjiang Province. The official investigation showed that the owner of the illegitimate (without any approval or permit or certificate) coalmine was Peng Guocai, Vice Director of Administration of Work Safety of District Taoshan in Qitaihe. According to law, an official in charge of work safety should not be allowed to be engaged in any production activity, but this coalmine owner had two offices: one in the District Office Building where he served as a Vice Director of work safety; the other near the coalmine where he had coalminers work in the pit without any safety protection equipment. On 7 August 2005, 123 coalminers were killed in another massive coalmine accident in Meixing, Guangdong. A news probe by CCTV-1 (the flagship terrestrial television channel of the China Central Television) showed that the coalmine was repeatedly ordered to terminate its production for its failure to be up to safety standards, but it was always passed through the examination or inspection by some government agencies because many officials in these agencies received the bribery offered by the coalmine owner.[3] The two coalmine accidents reveal that tacit backing of the government is the underlying reason for collusion between officials and coalminers.

Take for instance illegal use of land and tax fraud. In 2002, Greentown China Holding Ltd. (Zhejiang) planned to build a five-star hotel which would cover an area of 10 hectares and a high-end luxury living quarter. Zhang Jiameng, the then Party Secretary of Zhoushan City abused his authority, specifically demanding some governmental agencies to give a green light to Greentown's application for land use. Under his direct intervention, Greentown evaded miscellaneous taxes worth of CNY 5 million and in return Greentown sold a house with a floor space of 203 square meters at CNY 5, 100 per square meter, half of the market price, a disguised bribe of CNY 1.17 million.[4] In this case, the party secretary colluded with the real estate enterprise not purely for local economic development, but for both the improvement of public services and personal gains.

In addition to our theoretical intuition and specific cases to demonstrate the relationship between collusion and corruption, we can also perceive it from the government-firm collusion model in Chap. 2. Let us make a slight modification of the model. Suppose the punishment that an official participating in collusion expected from the central government is $p\rho F_s$, among which p stands for the accident probability, ρ for the probability of disclosure by mass media, F_s for punishment by the central government. The expected revenue of the official is $\zeta k \Delta c$, among which ζ denotes coefficient of transaction fees of collusion. This coefficient reflects the difficulty level of collusion: the less the value of ζ, the higher the transaction fees of collusion. In reality, due to supervision, anticorruption, disclosure by mass media and other factors, the conveyance of interests between local government and enterprises will result in deadweight loss. k stands for the proportion of quasi-rent that an enterprise can share with the local government; Δc represents quasi-rent, i.e., the cost saved by the firm if it chooses a bad mode of production. The incentive-compatibility constraint (ICC) that an official participates in collusion is as follows:

$$p\rho F_s \geq \zeta k \Delta c \tag{7.1}$$

According to the analysis in Chap. 2, we know the ICC (7.1) is tight. This implies that in the case of the occurrence of an accident and the probability of disclosure by the mass media, the greater the punishment from the central government, the higher the cost of transfer payment (bribery), thus the more serious corruption. When the central government predicts corruption will become serious, it will intensify punishment to prevent collusion. Thus, we come to the following corollary:

Corollary 7.1: Ceteris paribus, corruption in a particular region is in positive correlation with collusion.

7.1.2 Collusion and Corruption: An Empirical Analysis

In this section, we will provide a simple econometric analysis. Though we can find indicators to measure the level of collusion in the particular fields such as coalmine accidents, illegal use of land and tax frauds respectively, it is still currently difficult to find an indicator that can measure the comprehensive level of collusion in a region. In Chap. 6, we set the proportion of state-owned shares of a state-owned enterprise as a proxy variable of the level of collusion based on a sound assumption that state-owned enterprises are closely associated with the government in terms of both administration and capital. Therefore, it is comparatively easier for a state-owned enterprise to collude with the government, or the cost of collusion is much less costly. Now we expand this method to the level of regions, i.e., use the proportion of state-owned economy as a proxy variable to measure the level of collusion in a particular province, prefecture or a county.[5] Thus, we choose the proportion of the sales volume of all state-owned enterprises to the total sales volume of all enterprises temporarily for our specific purpose. To be specific, on the basis of using the 1999–2007 Chinese Industrial Enterprise Database, we compute the proportion of the annual sales volume of all state-owned enterprises to the total sales volume of all enterprises and set it as the proxy variable of the level of collusion.

Another problem we face is: how to measure collusion? Since collusion is furtive and illicit, it is difficult to measure (Banerjee et al. 2012). Currently, we have four methods to measure collusion (Nie and Jia 2011). The first is corruption perception index (CPI) published by Transparency International, which annually ranks countries on a scale from 100 (very clean) to 0 (highly corrupt) by their perceived levels of corruption, as determined by assessments of experts and merchants. According to the 2012 ranking list, China ranked 80th place (five places lower than the previous year 2011) of 176 countries, scoring 39.[6] The second method is to make a questionnaire survey among enterprises and individuals about the costs of time and money when they deal with government departments and agencies, such as the enterprise survey from the World Bank. The third method is cross-checking based on data from a variety of sources. For instance, Fisman and Wei (2004) compared the volume of imports reported by mainland China and the volume of exports reported by Hong Kong and then estimated the amount of tax evasion on the part of mainland China. The fourth method is to make a comparatively objective estimation of governmental behaviors by observing relevant data. Here we

Table 7.1 The impact of collusion on corruption

Level of corruption	*Slope*	*Standard error*	*t*	*P* > |*t*|
Proportion of state-owned economy	1.198	0.406	2.95	0.03
Constant term	4.065	0.140	29.07	0.000

adopt the fourth one to estimate the level of corruption in 31 provinces, municipalities, and autonomous regions. To be exact, we take the number of corruption and embezzlement cases filed per ten thousand public servants to measure corruption in a particular province or municipality or autonomous region.[7] The data about corruption and embezzlement cases filed are from *China Procuratorial Yearbooks* and the data about public servants are from *China Statistical Yearbooks.* Much literature in this field favors this method. For instance, Glaeser and Saks (2006) used a data set of federal corruption convictions in the USA to investigate the causes and consequences of corruption.[8]

We construct the 1997–2007 panel data at the provincial level and take "the number of corruption and embezzlement cases filed" as a dependent variable of the level of corruption, the level of collusion measured by the proportion of the sales volume of state-owned enterprises to the regression of these data. Theoretically, the level of corruption will in turn affect the proportion of state-owned economy; therefore, in order to mitigate this reverse causality, we set the explanatory variable lagged for one period. The regression results are shown in Table 7.1.

We do not put other explanatory and control variables in the econometric equation because our purpose here is not to reveal the causes of corruption but to show that collusion affects the level of corruption. Table 7.1 shows that the proportion of state-owned economy that represents the level of collusion at the level of 1%is significantly positive, i.e., the level of collusion is highly in positive correlation with corruption.

7.2 Collusion Aggravating Government Soft Budget Constraints

7.2.1 Soft Budget Constraints

In a normal market economy, if outputs of an enterprise are much less than its inputs, the enterprise will immediately cease production, or declare it bankrupt, or liquidate its assets. This is budget constraint. To be simplistic, market economy cannot guarantee the success of all enterprises, and some of them will be knocked out due to inefficiency. Janos Kornail (1979), a Hungarian economist, observed that in the 1970s many Hungarian money-losing state-owned enterprises could still survive with the support of government subsidies and bank loans. He termed this phenomenon as soft budget constraint. Originally, this term was regarded as one of the main characteristics of socialist countries in the shortage period, but now it has been widely used in economics research on all transition countries and even on advanced countries (Kornail 2003).

In a perfect market economy where the government will never be partial to the minority of enterprises, soft budget constraint will never occur because the function of the government is to serve all enterprises and provide public service. However, in a collusion environment, soft budget constraint is inevitable. When a local government colludes with an enterprise, the interests of the two sides are closely bound. This is reflected in the following three aspects: (1) by choosing a bad mode of production over a good one, the enterprise can help the local government realize a quicker economic growth and a more superficially lustrous political performance; (2) since the enterprise solves the unemployment of many local residents including some officials' relatives, it can even hold up the local government, demanding all forms of preferential treatment; (3) since the enterprise gets approval of choosing a bad mode of production by offering bribery to the local government, the two sides form a coalition of interests where the achievements of either party will do credit to the other or the failures of either party will bring humiliations to the other. Such being the case, even if the enterprise is on the verge of bankruptcy, the local government will support it; otherwise its own interests will be damaged.

The preferential treatments which the local government gives to an enterprise participating in collusion are as follows: (1) demanding local state-owned banks or commercial banks to provide the enterprise with loans, or delay pressing on it for repayment; (2) subsidizing the enterprise with financial funding; (3) helping the enterprise strip off bad assets; (4) arranging for capital injection, asset restructuring, or merger and acquisition by a lucrative enterprise; (5) demanding debt relief from banks; (6) demanding tax relief from taxing authorities; (7) nationalizing the enterprise by pump-priming it directly through State-owned Assets

Supervision & Administration Commission; (8) repurchasing stocks of the enterprise to push up their prices in the secondary market; (9) taking advantage of the loophole opportunities to help the enterprise apply for new projects and subsidies.

7.2.2 *A Case Study: A Tragedy of Photovoltaic Industry*

When systematic empirical evidence is unavailable, a case study is a sub-optimal choice for our purpose to illustrate the behaviors of local governments in collusion. In many cases about collusion in China, no other industry might be more reflective of the close relationship between local governments and enterprises than photovoltaic industry. Here we present a case to illustrate the relationship between the city government of Dayu, Jiangxi Province, and LDK Solar Hi-Tech Co., Ltd.[9]

LDK Solar Hi-Tech Co., Ltd., once the largest producer of polycrystalline solar wafers in Asia, is a joint venture with registered capital of $111 million and total investment of $300 million in High and New Technology Industrial Development Zone in Dayu, a city in Central China's Jiangxi Province. It was put into production in April 2006, and its capacity reached 100 megawatts in July and doubled in October, thus winning the epithet of "LDK growth wonder" and the title of "most emerging-growth enterprise in new material industry". In June 2007, it was successfully listed in New York Stock Exchange, making it the largest IPO ever in the history of China's new energy industry. However, this elite enterprise, under the pressures of explosive growth of photovoltaic enterprises in China and heavy countervailing duties and antidumping duties levied by the US government and the European Union, soon found itself riding the fading momentum. The annual reports from 2007 to 2011 showed that its total assets decreased year by year and its total liabilities increased from $617 million to $6010 million with an astonishing annual growth rate of 147.27%. Since 2010, its liabilities continued to swell to the extent that its creditors China Development Bank, Bank of China, and China Construction Bank, and others formed a special commission to supervise its financial affairs.

How did local governments involve themselves and banks in the production activities of LDK Solar through soft budget constraint?

In 2005, Peng Xiaofeng, founder and CEO of LDK Solar, originally planned to establish a crystalline solar wafer base in Suzhou, a city lies somewhere between Shanghai and Nanjing, but later he was persuaded by Wang Dehe, Party Secretary of Xinyu, into founding the base in Dayu, a city not far away from Peng's birthplace in Jiangxi. In order to attract this promising project which might propel local economy and promote its political performance, the Xinyu city government spared no efforts, investing in advance CNY 200 million, one-ninth of its total fiscal revenue of its previous year. The party secretary himself and a group of officials whom he convoked paid a visit to Jiangxi International Trust and Investment Company and got a loan of CNY 100 million; the remaining CNY 100 million was financed by Xinyu Finance Bureau and Jiangxi Finance Department.

LDK Solar also helped boost the local economy. In 2011, it paid tax of CNY 1360 million, the largest contributor to the financial revenue of Xinyu. Meanwhile, a massive industrial cluster with upstream and downstream firms was formed around LDK Solar. In 2011, the added value of the whole industry increased by 28.8% to CNY 12,016 million, and accounted for 33.9% of manufacturing industry. It was estimated that its contribution rate to the local economic development was 49.6%. LDK Solar became the second largest taxpayer in Jiangxi Province.

However, the Xinyu city government paid a heavy price for its involvement. When LDK Solar was heavily laden with debts, the Xinyu city government tried all conceivable means to help it with debt repayments. In the early of May 2015, at the request of the Jiangxi provincial government, the Jiangxi branch of China Development Bank, and other banks provided LDK Solar with a syndicated loan of CNY 2000 million. On July 12, 2012, the Xinyu Standing Committee of People's Congress convened its seventh conference, specially considering and approving an initiative to demand that the city government should incorporate into its annual fiscal budget LDK Solar's shortfall of the trusted loan to be repaid to Huarong International Trust Co,. Ltd., to which LDK Solar still had owed CNY 755 million by the end of 2010. In October 2012, the Xinyu city government stepped up its efforts to bail out the moribund giant by declaring that all enterprises with 60% of state-owned shares in Xinyu would purchase 19.9% of LDK Solar's shares in an attempt to nationalize it.

Fearing that photovoltaic industry would cause environment disasters, the local residents were not in favor of and even became hostile to LDK Solar. Due to fear of pollution, the price of the residential houses in the development zone or neighboring residential quarters dropped all the way down to CNY 1800 per square meter from the peak of CNY

3000. Besides, for many times some foremen and construction workers appealed to local authorities for project payment default. It is reported that the Xinyu city government had to spend CNY 300 million for stability maintenance. Ironically, even in the severe recession of 2009, LDK Solar did not lay off a single worker, but now it has to cut off 10,000 jobs.

What does the above case reveal? First, the relationship between LDK Solar and the Xinyu government is a sort of collusion. The officials (party secretary and the members of the standing committee of the local people's congress) of the local Xinyu government supported an enterprise with a "bad" mode of production—long production cycle, heavy pollution, and difficulty in coordinating supply and demand[10]—in the hope of easing the pressure of unemployment, increasing fiscal revenue and promoting its political achievements. Although the photovoltaic industry contributed a great deal to fiscal revenue, it also brought about negative social costs—such as social instability arising from LDK Solar's defaulting on project payments, which would be eventually passed onto the Central Government or the provincial government of Jiangxi.

Second, the repeated and continuous bailing out of a debt-ridden and insolvent enterprise is a typical matter of soft budget constraint. In the normal situation of a market economy, LDK Solar should make a declaration of bankruptcy. However, the Xinyu city government tried every possible means—such as arrangement of bank loan, financial guarantee, and pump-priming—to bail the enterprise out because it realized that bankruptcy would bring legions of laid-off workers, lower local fiscal revenue, and thus roil social stability. If LDK Solar fails to overcome difficulties, the real loss resulting from soft budget constraints will be eventually passed onto the taxpayers of Xinyu City and even of the whole of Jiangxi Province. Here the moral hazard is self-evident because the local government, as a special legal person, just undertakes limited liabilities.

7.3 Summary

This chapter discusses the impact of collusion on the behaviors of the local governments by focusing on corruption and soft budget constraint. The collusion between local governments and enterprises is in essence a covert violation of rules and regulations and it will inevitably exacerbate "trading power for money" and other judicial corruption. On the basis of a simple but effective empirical test of the relationship between corruption and collusion by using the 1997–2007 panel data at the provincial level, we have found that the level of corruption measured by the proportion of state-owned assets is in significantly positive correlation with the number of corruption and embezzlement cases filed. Next we have made an inference that collusion worsens soft budget constraint and made a case study of the photovoltaic industry to illustrate how local governments try to bail out an insolvent enterprise through the possible means of soft budget constraint.

The policy implications of this chapter are as follows: First, as collusion is likely to exacerbate the increasingly serious problem of corruption, the central government must take into consideration the channels through which how enterprises collude with local governments. Second, in the current social context of the currently high-profile "China Model" and "Powerful Government", it is as equally important to understand the drawbacks of the government as to understand the drawbacks of the market. If China intends to step toward a market economy ruled by law, it must sever the bonds of collusion and harden government's budget constraint.

Notes

1. In June 2003, Tieben, a private iron & steel company in Jiangsu province, illegally seized 400 hectares of arable land and began its inflated project boasting an annual output of 8 million tons of iron and steel without examination and approval by state departments concerned. More than 4000 peasants were forced to remove to other places, some of them having to live in shanties, bridge openings and deserted fishing boats. Soon, the State Council sent an investigative group and found out that the truth that some local government agencies and officials overstepped their authority to approve an illicit, inflated, polluting project. After the disclosure of the event, many officials were removed from office or brought to justice.
2. For instance, it is very common for a county (the third level of local government in China) to elect or appoint seven or eight vice-magistrates to be responsible for education, health care, science, culture, industry and so on. Bureaucracies are rampant in some counties where there are eleven or even more vice-magistrates. See "Who Appoints 50% More Vice-Magistrates?" http://news.liao1.com/newspage/2011/05/4502494.html (Accessed November 13, 2016 by the translator).
3. For full details of the news report and probe, refer to the TV programme of "Cut off the Hands Behind 'Collusion Between Officials and Coalmine Owners'", *Oriental Horizon*, 25 August 2005.

4. For full details of the news report and probe, refer to "Greentown CEO Greases the Palm of Zhoushan Party Secretary", *Yangzi Evening News*, January 18, 2012.
5. It must be stressed that this is a simplistic or crude indicator of the level of collusion. We hope our future research will provide indicators to measure more accurately the level of collusion at the level of administrative region (province, prefecture, county, and township).
6. Until 2011, the CPI scoring system had been based on the global score of 10. In the 2012 ranking list, Denmark, Finland and New Zealand shared the first place, Sweden the 4th place, Singapore 5th place, Australia 7th, Canada 9th, Germany 13th, Japan and Britain 17th, the US 19th, France 22nd. Link: http://cpi.transparency.org/cpi2012/results. (Accessed December 25, 2012).
7. Such a calculation method can remove the difference of the absolute number of public servants in different administrative regions. A province with a larger population and more subordinate administrative agencies has more public servants than a smaller counterpart.
8. Nie and Jia (2011) demonstrate the validity of this method and refute some accusations against it.
9. This case is collated by the author of the original from a series of news reports and probes heavily scripted in *Daily Information*, *China Business News*, *China Securities Journal*, and *China Economic Weekly*. When the Chinese original of the book was published in 2013, LDK solar was mired in debts and on the verge of bankruptcy. By March 2014, it was delisted from NYSE by the American securities regulator, the SEC (Securities and Exchange Commission).
10. Due to unchecked development of photovoltaic industry, 33 polycrystalline solar wafer bases had been in construction by March 2008 in 16 provinces and their annual production capacity was estimated to have reached 140,000 tons, far above the real global demand of 80,000 tons. In addition, Chinese photovoltaic enterprises had to rely on foreign countries for the importation of 90% of raw materials and the exportation of 90% of solar panels and cells. Therefore, it was difficult for a photovoltaic enterprise to coordinate its demand and supply.

References

Banerjee, Abhijit V., Rema Hanna and Sendhil Mullainathan. "Corruption". MIT working paper. 2012.

Fisman, Raymond and Shang-Jin Wei. "Tax Rates and Tax Evasion: Evidence from 'Missing Imports' in China". *Journal of Political Economy*. 2004, 112 (2): 471–496.

Glaeser, Edward and Raven Sacks. "Corruption in America". *Journal of Public Economics*. 2006, 90: 1053–1072.

Kornai, Janos, Eric Maskin and Gerard Roland. "Understanding the Soft Budget Constraint". *Journal of Economic Literature*. 2003, 41 (4): 1095–1136.

Kornail, Janos. "Resources-Constrained Versus Demanded-Constrained Systems". *Econometrica*. 1979, 47 (4): 801–819.

Nie, Huihua, and Ruixue Jia. "Institutional Quality and Ownerships of Foreign Direct Investment in China", Renmin University of China, working paper, 2011.

Svensson, Jacob. "Eight Questions about Corruption". *Journal of Economic Perspectives*. 2005, 19 (3): 19–42.

CHAPTER 8

The Impact of Collusion on Social Development

This chapter is mainly concerned about the impact of collusion on social development. Our major findings are: collusion does harm to supervision by public opinion, immobilizes residential income distribution pattern, hinders the development of social organizations, and increases aggregate cost of reform. Section 8.1 analyzes the relationship between collusion and public supervision; Sect. 8.2 discusses the impact of collusion on residential income distribution; Sect. 8.3 attempts to delve into the relationship between collusion and the development of social organizations; Sect. 8.4 discusses the impact of collusion on the costs of reform; Sect. 8.5 is a summary.

8.1 Harm to Supervision by Public Opinion

8.1.1 Supervision by Public Opinion and Collusion: Theory and Evidence

Supervision by public opinion, along with property rights protection, judiciary independence, and democratic election, constitutes the cornerstones of contemporary rule of law. It plays a robust role in collusion-proof implementation. In the triple game model of central government, local government, and firm, due to the limitation of manpower and other objective factors, the central government cannot be accurately informed of the behaviors of local governments and enterprises. Therefore, there exists information asymmetry between the central government and local

H. Nie, *Collusion, Local Governments and Development in China*,
DOI 10.1007/978-981-10-5059-6_8

governments and between the central government and enterprises. As mentioned in Chap. 2, an enterprise's choice of a bad mode of production means collusion where an accident (coalmine accident, pollution, and food adulteration) may happen at a certain degree of probability p and may be disclosed by the mass media at a certain degree of probability ρ. The central government knows the collusion between local governments and enterprises through the news coverage and probes by the mass media which, within our analytic framework, are the only source of information for the central government. According to Corollary 2.1, ceteris paribus, the higher the probability ρ, the less the sharing proportion distributed to the local government ($\frac{\partial \alpha^{cp}}{\partial \rho} < 0$), the higher the tax rate imposed on the enterprise ($\frac{\partial t^{cp}}{\partial \rho} > 0$). This is because the mass media's disclosure of collusion improves the central government's supervision and in this case, the central government will give less information rent to local governments and enterprises. As a matter of fact, according to the equation of (2.19), we know that the central government's utility level under the condition of robustly collusion-proof implementation is $\underline{U}^{cp}_{p} = 1 - \underline{c} - \frac{\Delta c}{p\rho}$. If we calculate the comparative static result of the utility level of the central government, we can get the following corollary:

Corollary 8.1 $\frac{\partial \underline{U}^{cp}_{p}}{\partial \rho} > 0$, *The improvement of supervision by public opinion is conducive to the improvement of the central government's utility level.*

Corollary 8.1 reveals a universal truth, i.e., the improvement of supervision by public opinion is advantageous to the central government. Here, supervision by public opinion includes not merely legacy media and network media (such as Microblog and Twitter) but also word of mouth and other forms of speech. In the real world, both case studies and econometric analysis can demonstrate that supervision by public opinion is the sword that can pierce collusion. The following case of illegal use of land can demonstrate the power of supervision by public opinion.[1]

On May 28, 2008, the Lingbao City Government of Henan Province "rented," in the name of establishing an industrial zone, 28 square kilometers of land, most of which was arable land owned by about 30,000 peasants. The city government announced compensation standards for buildings and crops on the land and decided to rent the land for a period of 30 years at a lump-sum rent of CNY 1000 per μ (approximately equal to 667 square meters). Some villagers studied the *Law of Land Administration* carefully and found that "rent replacing requisitioning" was an illicit behavior which was being cracked down by the central government, so they successively appealed to the higher authorities in Sanmenxia and Zhengzhou.[2] In order to mollify local villagers, the Lingbao City Government decided to raise the rent to CNY 1200 per μ and finally it reached an agreement with them. Soon, this news was spread to Wang Shuai, a migrant worker with a decent job in Shanghai, thousands of miles away from his hometown in Lingbao. Intuitively realizing the local government's abuse of authority and violation of rules, he logged into the websites of Sanmenxia Land and Resource Bureau for official documents about land management and learned that the Lingbao local government did not pay local villagers land compensation and settlement fees as stipulated by law. He made an online complaint against this illicit use of land with Henan Land and Resource Department. Not having received any reply, Wang had no alternative but to post some messages and pictures online in an attempt to arouse the attention from land departments concerned. This posting immediately grabbed the homepages of influential websites such as Sina, Sohu, and Yahoo and soon startled the Lingbao local government. On March 6, 2011, Wang was arrested by the police in Shanghai and was taken back to Henan in detention. Eight days later, his family reached an agreement with the police and Wang was bailed out. After the disclosure of this illicit use of land, the local government was reported to set up a standard of compensation settlement fees. Since this event caused a sensation through the Internet, the Henan Public Security Department was forced to withdraw the verdict of bail pending trial with restricted liberty of moving and made an apology to Wang and compensated him.

In the above case, the local government deliberately violated regulations and rented the land to establish an industrial zone—an ostentatious and unproductive project for many county-level cities—simply for the betterment of political achievements. This is a sort of collusion between the local government and enterprises that swindle development funds for false projects. It infringes the interests of the local peasants and has a negative effect on both the authority of the central government and the master plan for farmland protection. Ironically, normal appeal procedures could not pierce the strong shield of collusion alliance but only through network media could the peasants get back part of their own rights. This reflects not merely the value of mass media but also the social costs brought about by supervision by public opinion. The disclosure of an event or accident is likely to attract public attention and cause social instability, a social cost which will be eventually assumed by a higher local government or the central government. According to the collusion model in Chap.2, social instability is equated with the increment of social cost *C*, or the increase of the cost at which the central government permits collusion. In this case, the central government is prone to take collusion-proof measures.

The problem of environment pollution is a typical case of collusion. In this field, supervision by public opinion also plays an active role in fighting collusion. Can the press disclosure of environment pollution of an enterprise prompt it to make environment information known to the public? The answer is definitely affirmative. For instance, Aerts and Cormier (2009), by using the data of North American firms (including the United States and Canada), observe that negative media legitimacy is a driver of environmental press releases. Shen and Feng (2012) test the results by using the data of Chinese firms. They collect 206 sample heavy-pollution enterprises, all of which list their stocks in the Shanghai or Shenzhen Stock Exchange and were frequently mentioned in environmental press releases between 2008 and 2009. They sort out, via the China InfoBank, news reports about serious pollution and then classify them into positive, negative, and neutral media legitimacy. Negative media legitimacy here refers to the news coverage (excessive emission of pollutants or failure to be up to the environment standards) that will have a negative effect on the image of an enterprise. In compliance with *Measures for the Disclosure of Environmental Information* (for Trial Implementation) issued in February 2007, they divide the contents of the disclosures into six categories: (1) guidelines, annual objective, and achievements in environmental protection; (2) total annual consumption of resources; (3) investment in contamination control and technological development in pollution reduction, and the construction and operation of antipollution facilities; (4) types of environmental pollutants and their amount, concentration or potency, and whereabouts; (5) the expenditure on environment protection; (6) others. Each of the above categories is translated into a value. In their regression equation, the dependent variable is the level of environmental information disclosure, and the explanatory variable is the level of press supervision or disclosure of environmental problems. The econometric regressional results show that negative media legitimacy at the level of 1% significantly and positively improves the level of the disclosure of environmental information by the enterprise itself.

In the coal industry, Jia and Nie (2017), using the 1995–2005 coalmine accident and press data at the provincial level, measure the order of severity of coalmine accident with mortality rate (the number of coalmine workers killed in the production of per million tons of coal) and the level of press supervision with the number of newspapers (and other forms of mass media) per thousand people. They also find that press supervision significantly reduces coalmine accidents. Besides in the domains of land management, environmental protection, and work safety in coalmining, the press also plays an important role in food safety. For instance, in the Sanlu Milk Powder Scandal in 2008, it is the initial press release of "16 Gansu infants diagnosed with kidney stones due to Sanlu infant formulae" that prompted the National Bureau of Quality Inspection to immediately send an investigative group to Sanlu Group. In the subsequent scandals of melamine adulteration in dairy products, the press also plays an important role of supervision.

In summary, during the process of collusion-proof implementation, supervision by public opinion discloses collusion to the central government and exerts pressure on local governments and enterprises.

8.1.2 Collusion Obstructing Supervision by Public Opinion

While supervision by public opinion plays an important role in deterring collusion, it will be deliberately and craftily hindered by local governments and enterprises.

For local governments and enterprises, supervision by public opinion is, in the long run, is not exogenous and thus alterable under some special conditions. Therefore, they will use every possible unscrupulous means to sabotage supervision by public opinion: (1) forbid local mass media to report pernicious accidents; (2) grease the palm of news reporters with hush money; (3) bribe mass media by giving more advertising fees; (4) refuse to cooperate with correspondents; (5) delete unfavorable comments posted in the internet through their public diplomacy and social network.

A Case Study: How did local governments and the enterprise obstruct investigation in the Nandan Coalmine Accident? How did the central government inflict punishments on local officials for their negligence and deliberate suppression of the truth?

The pernicious Nandan Coalmine Accident reveals profoundly the confrontation between collusion and media superintendence. It happened in Guangxi Zhuang Autonomous Region in 2001 and became the first coalmine accident disclosed by mass media. The following is an account of this disaster based on Cheng (2008) and Zhao (2004).

On July 17, 2001, a water inrush accident struck a coal pit in Nandan County in Guangxi Zhuang Autonomous Region, killing 81 workers. Instead of inflicting a heavy punishment on the coalmine owner Li Dongming, the major governmental officials in Nandan colluded maliciously with him to cover up this serious dereliction of duty. Several days later, some bits or fragments of information about this terrible accident were spread among the folks via message boards of some popular websites such as 163.com, sohu.com, and Dayoo.com and quickly commanded great concern from many correspondents. It was until July 27 (ten days after the breakout of the accident) that a townsman provided clues to *The Nanning Evening Newspaper*, which immediately joined its subordinate *Bagui Urban Daily* and sent four correspondents to the scene of the accident. In their interview, the four correspondents were repeatedly intimidated by a bunch of thugs gathered by the coalmine owner to take their lives and the news coverage was not released to the public. On July 30, the correspondents of *People's Daily* to Nanning (capital city of Guangxi Zhuang Autonomous Region) sent to its headquarters in Beijing the first document for internal reference—"An Urgent Report about Nandan Coalmine Accident in Nanning." The following day, these correspondents sent to people.cn a news report "The Bewildering Truth of Nandan Coalmine Accident," which was quickly spread by many other websites. On August 2, the correspondents of *People's Daily* to Nanning sent to the editorial board of the headquarters "A Second Urgent Report of Nandan Coalmine Accident." At the top of the first page of the report were marked "SPECIAL URGENCY! SPECIAL URGENCY! SPECIAL URGENCY!". The editorial board attached great importance to this report and immediately submitted it to the Central Government in the form of "information special report." The afternoon of that day the then Premier Zhu Rongji of the State Council made a written instruction, demanding to make an in-depth investigation to the bottom of the truth.

On the morning of August 3, the correspondents of the *People's Daily* to Nanning drove to the scene of the coalmine accident. Keenly conscious of lurking dangers, these correspondents asked some departments concerned to send them three plainclothes policemen to escort and accompany them all the way. It was the first and the only one time for these correspondents to be protected by police in their 20-year career of news interviewing.

As might be expected, the investigation did not go along well. All family members and relatives of the coalmine victims received a warning sent by the coalmine owner's stooges: forbidden to stay at home, contact strange people, and disclose any details about the accident and compensation. With the slow progression of the investigation, people.cn released day by day the names of 12 victims, and the truth gradually came to light. On August 4, *Yangcheng Evening News* devoted a whole page to release the results of the preliminary report and disclose deliberate suppression of the truth by local officials, shocking the whole nation. That evening, people.cn also issued a briefing about the accident.

Ironically, from July 19 to August 2, both the administrative office of Hechi Prefecture (which administratively governs Nandan County) and the government of the Guangxi Zhuang Autonomous Region sent their working groups to investigate the truth for four times, but their reports showed "no casualties in the accident."

Later, the truth was eventually brought into daylight, and major local officials liable for negligence of duty received their punishment. Wang Hanmin, Vice-chairman of the People's Government of Guangxi Zhuang Autonomous Region, was given a recording of an administrative demerit for his misconducts in dealing with the accident. Mo Zhenhan, Party Secretary of Hechi Prefecture was sentenced to 14 years in prison for taking bribery of CNY 170,000 and $14,000, which was then turned over to the State Treasury. Yan Zhihua, Special Commissioner of Hechi Prefecture was sentenced to 1 year in prison with 2 years of reprieve for the negligence of duty. Wan Ruizhong, Party Secretary of Nandan County was sentenced to death confiscation of his personal property and illicit income worth of CNY 3,185,000 for taking bribery and abusing authority. Tang Yusheng, Magistrate of Nandan County, was sentenced to 20-year imprisonment, 3-year disfranchisement, and confiscation of personal property of CNY 200,000 and illicit income of CNY 360,000 and USD 1000 for taking bribery, embezzling public funds, and abusing authority. Mo Zhuanglong, Vice Party Secretary of Nandan County, was sentenced to 10 years in prison and confiscation of his personal property CNY 30,000 and illicit income of CNY 70,000. Wei Xueguang, Vice Magistrate of Nandan County, was sentenced to 13 years in prison and confiscation of his personal property of CNY 40,000 and illicit income of CNY 100,000. The defendant, Li Dongming was sentenced to 25 years in prison and a fine of CNY 200,000 for illegal mining.

Nandan Coalmine Accident is a disaster that should be ascribed to the powerful protection umbrella of local officials, including the party secretaries at the three local levels of the autonomous region, the prefecture and the county down to magistrate and vice magistrate at the county level. The case reveals that, in the campaign of fighting collusion, there is a tough backlash, such as blockading the passage of information. Under the intimidation of taking their lives, local correspondents were hushed. The truth was revealed only through message boards in the internet and mass media from other provinces. "Supervision by non-local mass media" reflects the fragility of the environment of supervision by public opinion in China.

8.1.2.1 The Predicament of Supervision by Public Opinion in China

Nandan Coalmine Accident is only one of the few cases that have been fortunately and successfully disclosed by mass media to the public. More often than not, supervision competence is very limited before the onslaught of powerful collusion. On the most fundamental level, the fragility of supervision by public opinion stems from the limitation of the governance system. In modern China, mass media are strictly under the control of the government. In particular, traditional mass media such as newspapers, periodicals, and magazines, TV and radio stations are all under the charge of a propaganda department affiliated to a party committee at different administrative levels. In the case in discussion, the local TV and radio stations and newspapers are closely censored by the propaganda department affiliated to Party Committee of Nandan County and the top officials in the department and committee take bribes to give a green light to illicit coalmine owners. Supervision by public opinion, if not protected by law, will be greatly discounted.

Throughout the world, it is universally acknowledged that the government should provide legal protection for freedom of the media and their right of supervision. As early as in 1776, Sweden established *Freedom of the Press Act*, the world's first constitutional protection of freedom of the press. Similarly, Colombia and Finland also established similar acts respectively in 1888 and 1919 (Zhan 2005). The formulation of the law concerning freedom of the press guaranteed press and public surveillance and thus reduced corruption and other social problems.

Although the Chinese government has issued some regulations or rules about press supervision and citizens' right to know the truth, it has not established a special act concerning freedom of the press. Such being the case, media surveillance cannot bring into full play in supervising law enforcement, promoting legislation, and fighting collusion. Many correspondents have to make a compromise when being intimated by powerful local governments and their underlings. Recent years frequently witnessed some provincial governments send their policemen to other provinces to arrest some of those correspondents who, the local governments believed, had ever published unfavorable reports that damaged their "reputation and image." As a matter of fact, the Constitution bestows each citizen with freedom of speech and the right of supervision and the right of supervision by the mass media and public opinion should be regarded as a natural extension of citizens' right to express their opinions and to know the truth (Yin and Yin 2006).

In our political-economic model of government–firm collusion, the press is the major source from which the central government obtains information. This fact effectually provides a policy implication for protecting the press supervision from the perspective of incentives. Even if we merely take into consideration the "technicality demand" for the central government's overseeing local governments and enterprises, the protection of press supervision and acceleration of news legislation will be still of great importance. Egorov (2009) provided us an interesting finding that the monarchs in many Middle Eastern dictatorial and tyrannical countries would also allow partially or almost free mass media to supervise officials so as to reduce moral hazards. Furthermore, Egorov et al. (2009) makes a regression analysis of the 1993–2007 panel data of about 150 countries finds that the importance of media freedom varies with the natural-resource endowment and that media supervision in resource-poorer countries seems to be much more important than in resource-richer countries. In this sense, there seems to be a substitution relation between the economic incentives given by the ruler to local officials and press supervision.

8.2 Collusion and Income Distribution

8.2.1 Collusion Affecting Channels of Income Distribution Gap

In recent years, income distribution gap has become one of the most serious social problems in China. Income distribution is reflected in the following two aspects: widening distribution gap; decreasing proportion of labor reward in national income. We hold that the two aspects are both influenced by collusion.

Table 8.1 The Gini indices of some major countries

Country	*Survey year*	*Gini index*
Sweden	2000	0.25
Japan	1993	0.25
Norway	2000	0.27
Canada	2000	0.33
France	1994	0.31
Australia	1994	0.32
UK	1999	0.34
USA	2000	0.38
P.R. China	2001	0.45

Source The IBRD's *World Development Report 2006: Equity and development* (pp. 280–281)

Let us first discuss income distribution gap, which is principally measured by Gini index.[3] According to the international standard, any value between 0.4 and 0.6 expresses a wider gap and a value above 0.6 expresses a tremendous gap. From Table 8.1, we note that China's Gini index (0.45) is higher than that of Northern European countries such as Sweden (0.25) and Norway (0.26) and some major industrialized countries such as Japan (0.25), Canada (0.32), France (0.33), Australia (0.35), and the UK. (0.36) except for the USA (0.41), indicating that China is among the countries where income distribution gap is seriously wide.

Since 2001, the China National Statistical Bureau has discontinued releasing its Gini index. Li Shi (2011), an expert on income distribution, estimated that China's Gini index in 2011 was about 0.52, higher than the figure listed in Table 8.1. According to the China Household Finance Survey released by Southwestern University of Finance and Economics in Chengdu, Sichuan, the Gini index of Chinese households was 0.61 in 2010, the urban intra-household Gini index 0.56 and the rural intra-household Gini index was 0.60. The Survey arrived at the following conclusion: "The income distribution gap between Chinese families is tremendously wide and it is rare around the globe."[4]

Many factors contribute to the widening distribution gap between Chinese families: gap between urban and rural areas, monopoly of industries, tax policy, education level, income liquidity, and corruption (Li 2011). However, up to now, there has been no literature to explore income distribution gap from the perspective of collusion, which actually has tremendously adverse effects on income distribution gap in the following two aspects.

First, collusion worsens corruption and embezzlement. Corruption, a tacit and illicit exchange of money for loosening supervision on corporate operations, is a major contributor to widening the income distribution gap. Imagine, where is the money that should be otherwise inputted into improving workers' wages and work safety? It flows into the corporate owners' personal pockets or is offered to local officials as bribes. In the previous case studies, it can be easily noted that many local officials collude to retain for their own use a large sum of money that is required by the central government to be distributed among peasants whose land is requisitioned for the construction of public facilities or industrial zones. In addition, some chiefs of a township, a village may collude with a higher administrative organ to embezzle and share a percentage of poverty-relief funds, or funds earmarked for improving medical services or other purposes. That is the reason why local officials are often caught flaunting their "wealth" by purchasing luxury cars and decent houses and then sued for their corruption and embezzlement. Therefore, collusion worsens corruption and embezzlement (Corollary 7.1), which further widens the income distribution gap. It should also be stressed that corruption may emerge not only between an individual and an official but also between an enterprise and a governmental organ. According to *The Criminal Law of the People's Republic of China*, any unit, state-owned or non-state-owned, will be convicted of unit offering bribes if it offers money or other forms of benefits to a governmental official in the course of collusion and any governmental organ that receives money or others forms of benefits from any enterprise will be convicted of unit taking bribes.[5] In China, various kinds of bribery aggravate income distribution gap.

Second, collusion worsens monopoly. In the case of monopoly that is thoroughly opposite to perfect competition, those enterprises participating in collusion will be protected by local governments in terms of factor cost and market expansion and thus form an irresistible force of monopoly. Monopolistic enterprises earn much more profits, which will be transformed into wages, bonuses, perks, and subsidies to be distributed among top executives, managers, and employees. In other words, monopoly widens primary distribution gap.

8.2.2 *Collusion, Extensive Mode of Growth and Labor Income Shares (LIS)*

The decline of the labor share of national income is another problem which has aroused wide public concern in recent years. In modern market economy, the major parties participating in income distribution are governments, residents, and enterprises which respectively are entitled to net production taxes, labor rewards, and earning surpluses. Therefore, the ratio of residential income to GDP can measure the ratio of labor reward to net national income. To be specific, lower labor income share indicates that higher shares of national income will be distributed among governments and enterprises. In other words, ordinary working residents earn much less than governmental employees and business operators do.

If calculated on the basis of national income accounting, the labor rewards for ordinary residents dropped by 9.75% from 49.49% in 1993 to 39.74% in 2007 while the net production taxes for governmental departments increased by 3.13% from 11.68 to 14.81% and the depreciation of fixed assets and earning surpluses (i.e., capital gain) for enterprises increased by 6.62 from 38.83 to 45.45% within the same period. Comparatively speaking, labor income share was too low and had a substantial drop.[6]

On the basis of the data of "Cash Flow Statement (in Kind)" in *China Statistical Yearbook*, Liang (2012) calculated the 1992–2008 ratios (see Fig. 8.1) of labor reward to net national income (excluding depreciation

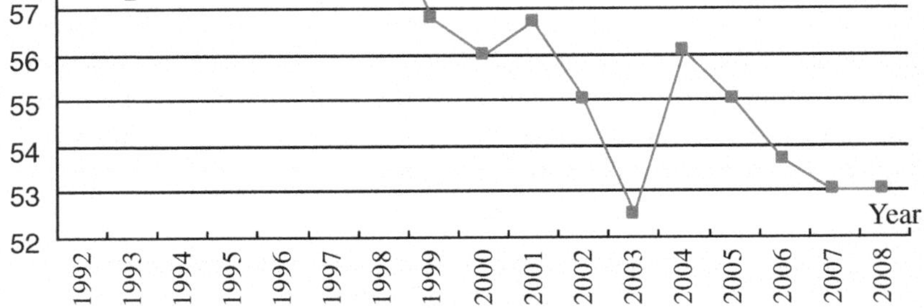

Fig. 8.1 Ratios of labor rewards to net national income (1992–2008)

of fixed assets). Figure 8.1 reveals that since 1992 the Chinese labor reward share has been dropping all the way to the current 50%.

Industrial structuring is often blamed for the substantial drop of labor income share. For instance, it is contended that the key factor consists in the drop of the share of agriculture and that the absolute value of labor income share in agriculture is absolutely higher than in other industries and services.[7] Granted, China's agrarian population share is large, but those migrant workers working in workshops still get a low pay. I hold the view that it is the extensive mode of growth motivated by collusion that widens the income gap.

To a large extent, the powerful Chinese economic engine is propelled by the extensive mode of growth featuring high input and output, heavy pollution, and low technology. This mode of growth is actually a "bad" mode of production motivated by collusion and its advantages rest only with cheap factors of production such as labor and land but not with technological innovation and product updates. Reflected in the primary income distribution, workers and land-lost peasants representing the owners of the cheap factors of production can only get a small part of profits and business operators and local governments enjoy the larger part of it. On the opposite, if an enterprise adopts a good mode of production propelled by technological innovation, skillful workers, technicians, and top executives could get a larger share in the income distribution pattern. Therefore, improvement of labor reward share and bridging the gap between rich and poor entails containing collusion and encouraging enterprises and local governments to follow a development road characterized with high technology, high quality, and environmental friendliness.

8.3 Collusion Hindering the Development of Social Organizations

8.3.1 *The Value of Non-governmental Organization*

The organization is one of the fundamental constituents of society. On the basis of the thinking of Williamson (1991), we classify all organizations into four categories: market, enterprise, hybrid, and government. Hybrid here refers to organizations such as nonprofit organizations (NPO), nongovernmental organizations (NGO), enterprise alliances, and franchise organizations. Each type of organization corresponds to a different contract with a different transaction cost and therefore each needs a particular governance approach which can by no means be arbitrarily substituted so that each of them can perform its own functions well in a harmonious society.

Seen from the perspective of politics and sociology, the development trend of a modern society is "small government and big society," otherwise known as "civil society". On the contrary, a society where the government strengthens and maximizes its own activities and solves all problems with governmental means is an extreme political society (Chatterjee 2004). In short, the development trend of a society is to let government, society, and individual give a full play to their own functions.

From the angle of development trend, enterprise, market, and government have had a widespread development throughout the world over the long history, but there is an exception to hybrid, especially to NGOs which have not yet fully developed since the concept was put forward by the United Nations in 1945 (Wang and Liu 2007). NGO is defined as any non-profit, voluntary citizen's group which is organized on a local, national, or international level. Like NPOs, NGOs are highly diverse groups of organizations including industry associations, charitable organizations, voluntary organizations, foundations, and societies.

NGOs are engaged in a wide range of activities which the government may not attach due importance to or which the government might thoroughly neglect or which the government is not good at. The roles of NGOs can be summed up as follows. First, they mobilize not only social resources including collecting charitable and social donations through all kinds of activities but also volunteers to be engaged in public welfare activities. Second, they provide public services and meet the demand of communities and residents for some public goods, such as various kinds of emergency aids in the 2008 Wenchuan earthquake. Third, they are engaged in social governance, including conveying public opinions to a governmental organ and defending civil rights. Fourth, they also play a major role in advocating the establishment of policies, including appealing to the government to legislate, supervising the enforcement of laws and promoting civic education.

8.3.2 Collusion: A Hindrance to Nongovernmental Organizations

NGOs play an irreplaceable role in varieties of social affairs, but in China, they have not yet been fully developed. According to *The 2009*

Civic Affairs Statistical Bulletin released by China National Ministry of Civil Affairs, by the end of 2009, there have been 425,000 registered social organizations in total, including 235,000 societies, 188,000 private non-enterprise units, and 1780 foundations. But ironically, most organizations are in effect official or semiofficial. Industry associations in many other countries are usually self-disciplined organizations voluntarily founded by enterprises or business owners; however, in China, they are mostly semiofficial and affiliated to an administrative unit which usually designates a retired governmental official to chair them.

Semiofficial and government-run industry associations can gain a lot of benefits from their close relationship with local governments: various kinds of support from government and maintenance of their reputation through heteronomy or a third-party. Nevertheless, they lack independence and thus find it difficult to defend their own vital interests and to maintain their reputation through self-discipline. Suppose an enterprise produces adulterated goods (adding melamine to infant milk formula). It severely damages the common interests of the whole industry and should be punished in accordance with the rules set forth by the industry association. If this enterprise keeps a close relationship with a local government, there will be nothing that the industry association can do about it. Other fellow enterprises within the industry association are very likely to follow suit and adulterate their products for more profits, and adulteration further degenerate into a prisoner's dilemma.

The above case gives us a revelation that collusion impedes the development of NGOs. Collusion stunts the growth NGOs through the following means.

First, collusion destroys the self-discipline of industry associations. Some enterprises in an industry association choose to collude with a local government for the purpose of getting support from local governments, which are de facto leader of the industry association. Therefore, the industry association can find no way to discipline those members colluding with the local governments.

Second, collusion suppresses the organizational ability of trade unions. If an enterprise colludes with local governments and chooses a bad mode of production that will endanger workers' health because of heavy pollutants, the trade union of the enterprise cannot come into its own being to bring into full play its role as a rights activist. Our theoretical prediction: the more serious collusion an enterprise gets involved in, the more difficult for its trade union to play a part in defending rights.

Third, collusion disrupts the operation of environment organizations. Collusion may cause serious environmental destruction and do harm to local residents' interests because the endeavors of environment organization to stop pollution will not take any desirable effect in the face of powerful local governments that serve as the umbrella of protection for that enterprise.

Fourth, collusion discourages enterprises from fulfilling corporate social responsibilities (CRS). In addition to being a working unit to maximize profits, an enterprise is also a social citizen who should fulfill some CSRs, including environment protection, community construction, and charitable activities. These CSRs, which also fall under non-governmental activities, will, in turn, promote the development of NGOs. However, under the adverse impact of collusion, the sole objective of an enterprise is to maximize profits at the cost of workers' health and of local residents' interests. Obviously, enterprises are reluctant to fulfill CSRs in the case of collusion.

8.4 Collusion Increasing Cost of Reform

8.4.1 Reasons for Early Reforms

Over the past three decades, China has made remarkable achievements through economic reform. Its gross domestic product has risen to the second place and per capita income has surpassed $4000 dollars and the population of poor people has shrunk by 2000 million. One of the reasons for its economic success is the adoption and implementation of the general strategic blueprint of "incremental reform preceding storage reform" (Fan 1993; Zhang 1996; Wu 2000), which can be simply put as reforming what is easy before reforming what is difficult.

For example, the household contract responsibility system launched in the rural countryside in 1982 helped peasants and collective units increase their income without impairing the interests of other communities. In the wake of the campaign of contract responsibility system was the proliferation and prosperity of township and village enterprises, which were actually the result of industrialization strategy in the rural countryside. These township and village enterprises are beneficial to the countryside because it resolved the problem of surplus workforce and enriched some people. Likewise, the emergence of special economic zones, coastal open cities, and costal economic development zones was also the result of the strategy line of incremental reform. Besides, the double-track system (coexistence of state-fixed price and the market price) in the mid-1980s also followed the incremental reform strategy.

The defining characteristic of incremental reform is that it improves the well-being of some people without doing harm to those with vested interests, and thus actualizing the economic principle of Pareto-improvement. Theoretically, Pareto optimality[8] is not merely an efficiency principle in economics but also a perfect moral standard.

8.4.2 Obstructions of Further Reforms

In the 1990s, China's reform was in full swing and brought about remarkable achievements, but it seemed to have come to a standstill over the past decade. In general, the reform has not yet accomplished its historic mission but entered a new critical stage of deepening reform. To do it justice, economic reform has been working well and has made stunning achievements, yet political reform has just started its first toddling step like a baby and social reform cries out for shouldering tough responsibilities in its long journey. As a matter of fact, China is now confronted with a lot of tough problems: housing; the development of agriculture, the countryside, and farmers; urban–rural segmentation; income distribution gap; education; social security; corruption; industry monopoly. Since storage reform is inevitable when reform enters the stage of deepening reform, it will unavoidably affect some people with vested interests and therefore Pareto-style incremental reform will find itself unsustainable (Nie 2011).

A further reform cannot be carried forward unless it is supported by those who have gained vested interests. This is the logic necessity of Pareto-style reform. However, if a group of people enjoys too many benefits from an existing economic or political privilege, it will be difficult to promote reform. And the next problem is: what decides the vested interests of those people? As far as I observe, collusion increases the vested interests of local governments and enterprises colluding with them and hence set barriers against the central government's top-down reform.

In order to reveal the logic relationship between collusion and the difficulty in propelling reform, I will explain it with Nash bargaining solution (Nash 1950). Suppose two-game subjects—the central government and a local government (or an enterprise)—negotiate on a reform plan. Given the reservation utilities (i.e., vested interests) of both parties are

$\bar{U}_1$ and $\bar{U}_2$ respectively, the aggregate interest after reform is U, and the interests of both parties after reform are U_1 and U_2 respectively. Then the reform plan with maximized aggregate interests of both parties is presented as the following mathematical programming:

$$U^* = \underset{U \in X, U \geq U}{\arg\max} \, (U_1 - \overline{U_1})(U_2 - \overline{U_2}) \tag{8.1}$$

Solving this programming, we can obtain a Nash bargaining solution[9]:

$$U_i = \bar{U}_i + \frac{1}{2}(U - \bar{U}) \tag{8.2}$$

The intuitive meaning of the Nash bargaining solution is: two agents in the negotiation obtain their own reservation utility plus 50% of net cooperative surplus. According to the formula (8.1), the higher one party's reservation utility is, the higher proceeds it can get in the negotiation. This implication is: suppose the aggregate proceeds are given in the reform negotiation between the central government and local governments (or enterprises), the more vested interests local governments and enterprises have obtained, the more difficult to push forward the reform.

In the absence of collusion, there is nothing special about the contract between local governments and enterprises, so both of them have neither too many vested interests nor overwhelming bargaining power. Thus, the cost of reform, for the central government, will be lower and the reform will be much easier to be pushed forward.

By contrast, collusion prompts local governments and enterprises to form a coalition of interests. Compared with the situation where there is no collusion, both parties have higher information rent and higher reservation utilities. In this case, both parties will ask for more incentive measures or more compensation for their vested interests because they are possessed of a stronger bargaining power when facing the reform plan proposed by the central government. This undoubtedly increases the cost of reform.

In addition, information asymmetry between the central government and local governments (and enterprises) will lead to an increase in cost from another channel. In the case of information symmetry, if the transaction cost is not high, both parties may come to terms about a Pareto-improvement. This is the well-known Coase theorem (Coase 1960). However, in the case of information asymmetry, a party with more information is inclined to conceal its own type and desirous of more information rent. If both parties attempt to do so, the information rent they attempt to ask for will surpass the aggregate net cooperative surplus, thus leading to a failure in coming to terms about a Pareto-efficient transaction. This is the famous Myerson-Satterthwaite theorem (Myerson and Satterthwaite 1983). Suppose the central government decides to introduce an energy-saving and emission-reduction program, local governments and enterprises will ask for compensation. Due to information asymmetry, both local governments and enterprises will be motivated to ask for more compensation than is really needed. Once the central government judges, after a careful consideration, that the compensation is too high, it will lay aside the reform plan that should otherwise be pushed forward.

8.5 Summary

This chapter discusses the social impact of collusion on many factors such as supervision by public opinion, income distribution gap, NGOs, and continuation of reforms. We have found that collusion counts against supervision by public opinion, ossifies current income distribution pattern, impedes the growth of NGOs, and increases social cost.

The policy implications of this chapter are as follows:

First, in order to contain collusion, the central government should jumpstart the legislation procedure of a press law to protect the legitimate interests of press supervision, especially to prevent local governments from inhibiting the freedom of speech of the mass media. Second, bridging income distribution gap should be accompanied with the suppression of all forms of corruption closely associated with collusion and with elimination of market monopoly under the local protectionism. Third, the central government should cut off the conveyance of interests between local governments and enterprises and construct a normal contractual relationship between them so as to protect the growth of NGOs. Fourth, the central government should mobilize the social communities to weigh in social reforms and check up costs and revenues of all social reforms instead of merely pushing top-down reform in case revenues of the reform might be soaked up by those groups with vested interests.

Notes

1 For full details of this report, see "An Internet Posting for Imprisonment of Eight Days", *China Youth Daily*, March 8, 2009.

2 Lingbao is a county-level city under the jurisdiction of Sanmenxia, a prefecture-level city in Henan Province, whose capital city is Zhengzhou.
3 Gini index or Gini coefficient is a concept proposed by the Italian statistician and sociologist Corrado Gini in his 1912 paper "Variability and Mutability". It is a measure of the degree of income distribution gap and its value ranges between 0 and 1. A Gini index of 0 expresses perfect equality while a Gini index of 1 expresses maximal inequality.
4 "Survey Report of Southwestern University of Finance and Economics: Gini index of Chinese household income tremendously high at 0.61", *The Beijing Times*, December 10, 2012. Link: http://sc.people.com.cn/n/2012/c346366-1783723.html (Accessed December 12, 2012).
5 Article 393 prescribes: Where a unit offers bribes for the purpose of securing illegitimate benefits or, in violation of State regulations, gives rebates or service charges to a state functionary, if the circumstances are serious, it shall be fined, and the persons who are directly in charge and the other persons who are directly responsible for the offence shall be sentenced to fixed-term imprisonment of not more than 5 years or criminal detention. Article 387 prescribes: where a state-organ, state-owned company, enterprise, institution or people's organization extorts from another person or illegally accepts another person's money or property in return for securing benefits for the person, if the circumstances are serious, it shall be fined, and the persons who are directly in charge and the other persons who are directly responsible for the offence shall be sentenced to fixed-term imprisonment of not more than 5 years or criminal detention. Any of the units mentioned in the preceding paragraph that, in economic activities, secretly accepts off-the-book rebates, or service charges of various descriptions shall be regarded as guilty of acceptance of bribes and punished in accordance with the provisions of the preceding paragraph.
6 Jia, Kang. "The share of labor rewards cries out for an objective analysis". *People's Daily*. May 18, 2010 (Accessed June 20, 2010).
7 Bai Chongen. "Why labor income share constantly falls?". *People's Daily*. May 17, 2010 (Accessed June 20, 2010).
8 Pareto efficiency, otherwise known as Pareto optimality, is named after an Italian economist and engineer Vilfredo Pareto (1848–1923), who used the concept in his studies of economic efficiency, income distribution, and allocation of economic resources. In an efficient state of allocation of economic resources, it is impossible to make any one individual better off without making at least one individual worse off.
9 Nash bargaining solution is usually a Shapley value, a solution concept of in cooperative game theory. See Shapley (1953).

References

Aerts, Walter and Denis Cormier. "Media Legitimacy and Corporate Environmental Communication". *Accounting, Organizations and Society*. 2009, 34 (1): 1–27.

Chatterjee, Partha. *The Politics of the Governed*. New York: Columbia University Press. 2004.

Cheng, Meidong. *A Perspective of Unexpected Massive Serious Emergencies in China Between 1949 and 2005*. Beijing: Zhonggong Dangshi Press (The Press of the History of the Communist Party of China). 2008. 程美东主编. 透视当代中国重大突发事件1949–2005. 北京: 中共党史出版社. 2008.

Coase, Ronald H. "The Problem of Social Cost". *Journal of Law and Economics*. 1960, 3: 1–44.

Egorov, Georgy, Swergei M. Guriev and Konstantin Sonin. "Why Resource-Poor Dictators Allow Free Media: A Theory and Evidence from Panel Data." *American Political Science Review*. 2009, 103 (4): 645–668.

Fan, Gang. "Two Reform Costs and Two Reform Approaches". *Economical Research Journal*. 1993 (1): 3–15. 樊纲. 两种改革成本与两种改革方式. 经济研究. 1993 (1): 3–15.

Jia, Ruixue, and Huihua Nie. "Decentralization, Collusion and Coalmine Deaths in China". *Review of Economics and Statistics*. 2017, 90 (1): 105–118.

Li, Shi. "Major Questions in Income Distribution of China". *Exploration and Free Views*. 2011 (4): 8–12. 李实. 中国收入分配中的几个主要问题. 探索与争鸣. 2011 (4): 8–12.

Liang, Ji. "An comparative analysis of labor reward share from a global perspective". *Review of Economic Research*. 2012 (45): 69–76. 梁季. 劳动报酬占比的国际比较与分析. 经济研究参考. 2012 (45): 69–76.

Myerson, Roger and Mark Satterthwaite. "Efficient Mechanism for Bilateral Trading". *Journal of Economic Theory*. 1983, 29 (2): 265–281.

Nash, John F., Jr. "The Bargaining Problem". *Econometrica*. 1950, 18 (2): 155–162.

Nie, Huihua. "The Rethinking of China's Deep-Seated Reform: From the Perspective of Incomplete Contract Theory". *International Economic Review*. 2011 (1): 129–140. 对中国深层次改革的思考: 不完全契约的视角. 国际经济评论. 2011 (1): 129–140.

Shapley, Lloyd. "A Value for n-person Games". In H. W. Kuhn and A. W. Tucker, eds. *Contributions to the Theory of Games*, vol. II. Princeton: Princeton University Press, 1953.

Shen, Hongtao and Jie Feng. "Supervision by Public Opinion, Governmental Surveillance, and Disclosure of Corporate Environmental Information". *Accounting Research*. 2012 (2): 72–78. 沈洪涛, 冯杰. 舆论监督、政府监管与企业环境信息披露. 会计研究. 2012 (2): 72–78.

Wang, Ming and Qiushi Liu. "An Institutional Analysis of the Development of China's NGOs". *China Non Profit Review*. 2007 (1): 102–155. 王名, 刘求实. 中国非政府组织发展的制度分析. 中国非赢利评论. 2007 (1): 102–155.

Williamson, Oliver. "Comparative Economic Organization: The Analysis of Discrete Structural Alternative". *Administrative Science Quarterly*. 1991, 36: 269–96.

Wu, Jinglian. "Retrospection and Prospect of China's Reforms". *Comparative Economic & Social Systems*. 2000 (2): 4–6. 吴敬琏. 中国改革的回顾与前瞻. 经济社会体制比较. 2002 (2): 4–6.

Yin, Chao and Yin Zenggang. "The Legal Ground of Supervision by News Media and Public Opinion: On News Legislation in China". *Youth Journalists*. 2006 (9): 59–61. 尹超, 尹增刚. 新闻舆论监督的法律依据: 兼论我国的新闻立法. 青年记者. 2006 (9): 59–61.

Zhan, Jiang. "A Global Overview of Legal Protection and Ethical Constraints of Supervision by Public Opinion". *Journal of China Youth College for Political Sciences*. 2005 (5): 121–126. 展江. 各国舆论监督的法律保障与伦理约束. 中国青年政治学院学报. 2005 (5): 121–126.

Zhang, Jun. "Why Can China's Incremental Reform Be Successful?" *Economist*. 1996 (3): 71–80.

Zhao, Shilong. *A Probe into China*. Beijing: China Founder Press. 2004. 赵世龙. 调查中国. 北京: 中国方正出版社. 2004.

CHAPTER 9

Institutional Environment of Collusion

This chapter attempts to make a research on the institutional environment of collusion so that we have an insight into and a solution to the problems arising from collusion. Section 9.1 analyzes the relationship among centralization, decentralization and collusion. Section 9.2 explains why there exist cyclical collusion and cyclical intervention. Section 9.3 discusses the law and policy environment of collusion. Section 9.4 researches the impact of regional competition on collusion. Section 9.5 is a summary.

9.1 Centralization, Decentralization and Collusion

9.1.1 Different Combinations of Political and Economic Systems

Institutions that create the main economic actors—human and physical capitals—is the foremost important factor of long-term growth (Acemoglue et al. 2005). A good understanding of the relationship between institutions and collusion is the prerequisite for a thorough understanding of the relationship between institutions and economic growth. In the previous chapters, we have discussed the game playing between the central government, local governments, and enterprises on the assumption that the institutional environment is given. In this section, we will loosen the assumption and discuss the impact of different institutions on collusion. From the perspective of administrative system proper, we can categorize all states (or economic bloc) into four types

H. Nie, *Collusion, Local Governments and Development in China*,
DOI 10.1007/978-981-10-5059-6_9

Fig. 9.1 Combination of political and economical powers

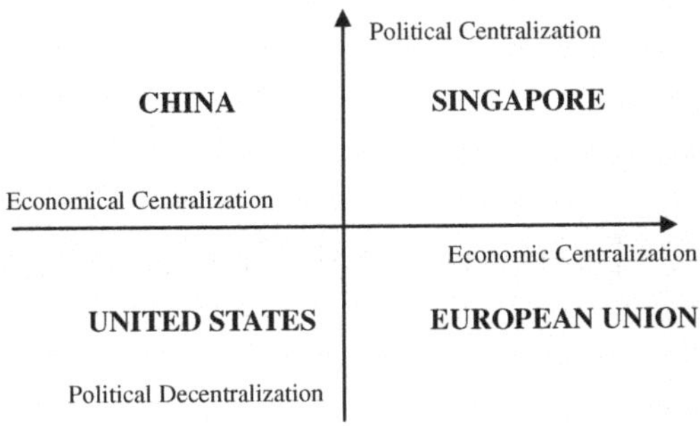

(see Fig. 9.1) based on the share of political and economical powers between the central government and local governments: politically and economically centralized states, i.e., absolutely centralized states; politically centralized and economically decentralized states; politically decentralized and economically centralized states; politically and economically decentralized states, i.e., absolutely decentralized states.

Under political centralization, the central government takes charge of national administrative affairs, including the exercise of jurisdiction over local governments and appointment and dismissal of local officials. Under economical centralization, the central government makes overall arrangement of economic affairs, including intervention in and decision-making of local economic development plan.[1] According to our classification, Singapore is an absolutely centralized state and its political and economic affairs are both decided by the central government. In the strict sense, Singapore has no local government, and it is divided into 5 districts, each of which has 87 constituencies. Only a few countries including some emirates fall under such a category. China belongs in the second category: politically centralized and economically decentralized state. From the perspective of the state system, China is a unitary parliamentary (congressional) republic where the central government (including the Central Committee of the Communist Party of China) takes charge of national important political affairs, and local governments must obey the command of the central government. On the other hand, a local government at any level, especially a provincial government, holds greater economic autonomy in the administrative area over which it has jurisdiction. In particular, since the implementation of tax-sharing system reform in 1994, there has been, in terms of finance and taxation, an asymmetrical decentralization relationship between the central government and local governments. This is fiscal federalism with Chinese characteristics (Qian and Weingast 1996). Nevertheless, it is not a pure fiscal federalism in that the central government and local governments do not decentralize their authority within a constitutional framework and furthermore they are by no means in peer-peer relation due to the fact that the central government has absolute political authority which will naturally endow them with more authority over economic decision-making.[2] After all, political authority decides economic system (Acemoglu et al. 2005). The third category is politically decentralized but economically centralized state. Such a state is rare. Now that politically it is decentralized, how could the central government decide national economic affairs? Obviously, there will not be such kind of a state unless a special law specifies that a special agency should be responsible for the coordination of economic affairs of different regions within this state. European Union falls within this category. On the one hand, European Union is a super state comprising 27 members, each of which has its comparatively independent political system with its own head of state or a chancellor or a premier or a prime minister and its own national defense and foreign affairs. On the other hand, European Union, as a common market, has a single currency (the euro), a single central bank (the European Central Bank), and a European parliament whose responsibility it is to coordinate economic affairs among its members. The fourth category is politically and economically decentralized state. This category includes advanced capitalist states such as the USA, the UK, Germany, and France (although both Germany and France are members of the EU).

9.1.2 Political Centralization: An Inevitable Contributor to Collusion

What different forms of collusion will be under different administrative systems? We still follow the assumption that there is only one central government (principal), a local government or a department directly under the central government, (supervisor) and an enterprise (agent). The central government faces a tradeoff between economic growth and social stability. Within our research framework, political centralization denotes that the local government should be responsible to the central government. Political decentralization denotes that the local government should be responsible for the local electorate (principal).

Economic centralization denotes that the central government directly controls the enterprise and decides what mode of production (good or bad) should be adopted. In the case of economic centralization, the local government or the department directly under the central government is unnecessary or they just provide the central government with some information about the enterprise but do not have any authority of decision-making over economic affairs. In other words, their information is only "cheap talk", which the central government may attend to or may neglect. Economic decentralization denotes that the local government is entrusted by the local electorate to supervise the enterprise. The local government can influence the enterprise's choice of mode of production, and moreover, the information it provides to the principal will affect the principal's decision-making.[3] According to the theory that political authority determines economic system, we can assume that, in the case of the coexistence of political centralization and economic decentralization, the de facto principal of the local government is the central government in terms of economic decision-making.

For the issue under discussion to be more interpretative, we assume that the loss (social cost) that collusion brings about to the principal (the central government and the local electorate) exceeds the revenue (economic growth). In this case, the principal is intended to take collusion-proof measures. Otherwise, the principal will always allow collusion. Besides, the government's revenue under centralization is higher than under decentralization because it may not necessarily share the revenues with the local government. This happens to be the classic conclusion of organization economics: with no limits on communication between the principal and the agents or no communication between agents, centralization always dominates at least weakly delegation (Laffont and Martimort 1995).

Let's discuss absolute centralization first. Under pure centralization where the central government functions as the principal and directly decides the production mode for the enterprise, there are no conditions to engender collusion for the following two reasons: the central government commands the enterprise to choose a "good" mode to defend social harmony; the local government, or a department directly under the central government cannot get any information rent due to the fact that local governmental officials are directly appointed by the central government. This corresponds with what is called "bureaucratic policy" by Tirole (1992) where collusion is prevented through relieving the local government of its authority. However, this form of collusion-proof strategy is costly in that, if compared with collusion, the enterprise will be less motivated to conduct production activities that might be highly efficient on the condition of much information rent, without which the government would also be less motivated to provide any valuable information. In other words, the purpose of centralization is to maintain social stability at the cost of reducing economic interests.

Next we discuss political centralization and economic decentralization, where the central government as the principal delegates its power to the local government to supervise the enterprise's production activities and where the local electorate as the principal of economic affairs has no authentic authority over the local government. Since the local government is the supervisor of the locality, its information structure is better than that of the central government but may be weaker than that of the enterprise; otherwise, the local government would be unnecessary. In the case of information asymmetry, the local government, if not responsible for the social costs of economic growth, has decision-making authority over the enterprise and affects the central government, thus inevitably giving rise to the incentive for collusion in which accidents may occur at a certain probability due to the choice of a bad mode of production. If the central government allows collusion, collusion is destined to happen. Such being the case, the central government cannot prohibits collusion unless it "redeems" the central government and the enterprise through economic interests. Therefore, in an economically decentralized state, the central government faces a tradeoff: either collusion at the sacrifice of social stability or deterrence of collusion at the cost of economic interests.

Now we come to political decentralization and economic centralization. In terms of production, the central government is not responsible for the local electorate, so it may allow the enterprise to choose a bad mode of production. But this is in collision with the interests of the local government because political decentralization denotes that the local government should be responsible for the local electorate, who are by no means willing to allow that bad mode of production. Here, the local government, as the administrator of the locality, may intervene in production activities without the constraints from the central government. Hence, under political decentralization, whether collusion might occur depends on the balance of power between the central government and the local government, for they have conflicts in economic interests.

Finally, we discuss political and economic decentralization. Under absolute decentralization, the local government is responsible, both economically and politically, for the local electorate, who are by no means willing to allow a bad mode of production. More importantly, the local electorate, as the participant of economic activities, knows well the enterprise's mode of production, so information asymmetry does not exist. In this case, the local government could find no way to seek illegitimate gains through an agency problem, so it has to ask the enterprise to choose a good mode of production. Therefore, collusion will never occur under absolute decentralization.

Let us relax some of our assumptions. After all, social governance is much more complicated than expected, and even a small state will exert absolutely centralized authority over its countrymen. A more pragmatic centralized administration is such where the central government delegates authority to a directly affiliated department to administer or govern the enterprise. Such an administration is an economic decentralization to some degree because the directly affiliated department, once having obtained authority from the central government and assuming no responsibility to the local electorate, tends to have the motivation to collude with the enterprise, the same consequence of political centralization plus economic decentralization.

Here is an interim summary: In any politically centralized state with a certain degree of economic decentralization, information asymmetry will be inevitable and local governments will be responsible for the central government but not for the local electorate, thus resulting in collusion. Comparatively, in the case of political decentralization where authority is bestowed to the local electorate to supervise local governments and enterprises, collusion will never occur in that information is perfect and symmetrical between the electorate and enterprises, and the electorate will be naturally reluctant to be the victim of a bad mode of production. A politically decentralized yet economically centralized state is not immune from collusion because there is conflict of interests. Of all the four categories of administrative systems, only a politically and economically decentralized state is thoroughly immune from collusion.

We further relax our assumptions. If allowing collusion may be beneficial to the central government and the local electorate, political decentralization is equivalent to political centralization. If considering interregional externalities, centralization will be much more advantageous to coordination and meanwhile to reduction of negative externalities arising from interregional competition. In this case, political centralization is superior to political decentralization.

Our final conclusion is: From the perspective of collusion deterrance, political decentralization plus economic decentralization (absolute decentralization) is optimal, followed by political decentralization plus economic centralization. Absolute centralization and political centralization plus economic decentralization are equally undesirable. From the perspective of allowing collusion, absolute centralization is optimal, followed by absolute decentralization. Therefore, for the central government, whichever is better (centralization or decentralization) depends on the central government's goal and its strategy toward collusion; whether collusion is permitted or not depends on its comparative research of the levels of well-being between allowing and deterring collusion.

9.2 Cyclical Collusion

In the collusion model in Chapter Two, we have discussed the central government's choices: allowing or deterring collusion. However, the real situation is much more complicated than this. The trade-off between economic growth and social stability is the political objective determined by centralized system, so the central government will not always persist in allowing or deterring collusion but make a choice between them at different stages according to different situations. Reflected in macroeconomic policies is the central government's cyclical intervention in collusion and cyclical collusion between local governments and enterprises.

Take coal mining. Once there is a serious coalmine accident in a particular region, the central government or a local government tends to adopt a one-size-fits-all safety check, or even order all coal mining enterprises to suspend production for rectification. For instance, the evidence by Xiao et al. (2011) shows that the Shanxi provincial government ordered all coalmines except key mining affairs bureaus to suspend production for rectification for one week in the wake of a series of coalmine accidents in August 2003; the Guangdong provincial government immediately ordered all 253 coal mining enterprises to suspend production for rectification after a serious water inrush accident struck a coal pit in Meizhou Guangdong in August 2005; all coal pits were ordered to be closed for rectification after a massive gas explosion hit a coalmine pit in Lengshuijiang in Hunan province in June 2005.

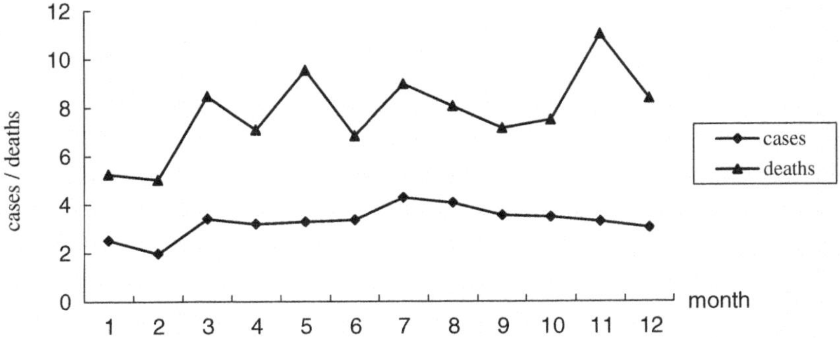

Fig. 9.2 Changes of number of coalmine accidents and deaths

Also in the coal mining sector, Nie et al. (2012) made an analysis of the 2000–2010 provincial-level monthly data and have an interesting finding: during the month when each province convenes its annual provincial congressional session or its annual session of political consultative conference, the number of coalmine accidents and death toll significantly reduce if compared with those in other months (see Fig. 9.2). They attribute this phenomenon to economic fluctuations driven by political cycles and dub it as political business cycle with Chinese characteristics.

What interests us is the following question: why does the central government periodically intervene in enterprises' business activities? Or is there a certain kind of cyclical collusion between local governments and enterprises? To explain this question, we insist that it is necessary to make an in-depth analysis of the cost and revenues that a bad mode of production brings to the central government. In the previous chapter, we have exogenously assumed that the social cost of per unit production is C, but we haven't analyzed other parameters such as output, probability of the occurrence of accidents, probability of disclosure by mass media, and the degree of tolerance of the public.

Nie and Zhang (2012) have answered the above question by expanding their collusion model. They have observed that the central government tends to allow collusion if the profits of economic growth exceeds the cost of social instability that may arise from all sorts of accidents. Besides, comparative statics shows that the product price (the advantage of economic growth), the probability of the occurrence of accidents and the public preference also have an impact on the central government's choice. As a matter of fact, these parameters may assume different values in different regions and at different stages. For instance, the public in a poor region may be more tolerant of environment pollution or prefer the advantage of economic growth, and therefore collusion in this region is more likely to be allowed. Even if in the same region, the situation differs before and after the occurrence of an accident. The cost of collusion before the occurrence of an accident is comparatively lower and the central government and local governments can tolerate it. However, after the disclosure of a massive accident by mass media and the simmering of public resentment against local governments' corruption or negligence of duty, the cost of collusion will be higher, and hence the central government and local governments will be on the side of collusion deterrence. This explains cyclical intervention in and rectification of the production activities of enterprises.

From the perspective of the central government or a higher local government (at the provincial and prefectural level), cyclical intervention is most probably ex post optimal. However, it is not necessarily so to enterprises, and especially disadvantageous to those enterprises adopting a good mode of production. Nie (2005) argues that cyclical intervention is actually a static game between the governments and unqualified enterprises. Due to serious economic loss and adverse social effect arising from the negligence of work safety on the part of unqualified enterprises, the optimal choice for the governments is to launch a one-size-fits-all campaign to suspend the business activities of similar enterprises across a region or even across the whole country. Unqualified enterprises, when predicting that governments will launch a new round of one-size-fits-all campaign, are less motivated to invest more money in the improvement of work safety coefficient, and even the qualified enterprises are more reluctant to invest in safety facilities because even the enterprise with safest safety facilities will be temporarily closed for rectification too. This campaign results in more safety accidents, and in turn the governments adopt more stringent measures against all enterprises of similar kind. In this cat-and-mouse fame, the governments and enterprises are both losers and its consequence is a vicious circle of safety accidents.

In order to avoid this vicious cycle, the governments must persists in *ex ante* policy optimality and consistency, or ex post punishment on enterprises involved in safety accidents and premium on enterprises without any accident for many years. However, the governments' pledge is unreliable because only the ex post optimal choice of a one-size-fits-all strategy can pacify the public resentment and win their support and understanding, thus reducing the cost of maintaining social stability. This is what is known as dynamic inconsistency of economic policies, a research domain to which Finn Kydland and Edward Prescott were awarded with the 2004 Nobel Memorial Prize for their contribution.

9.3 Policy Traps and Discretionary Power

In the end of 2001, I was making a field investigation in South China. At a symposium sponsored by a city government, a local official remarked: "I bet, if all Chinese enterprises pay 100% of taxes in accordance with the regulation, half of them will be closed". This thought-provoking remark contains a lot of information. First, local governments know the score but still wink at tax evasion. This reveals the existence of collusion. Second, a local government connives at tax evasion because it has the goods on enterprises and browbeats them into silence by asking them to do as it intends them to, such as demanding them to collude. Third, if the official got it just about right, why does the central government impose such a high tax rate on enterprises? A likely interpretation is that a higher tax rate helps the central government to compensate for a heavy loss due to tax evasion; otherwise, tax revenue will be much less. Another likely interpretation is that the governments at different levels need a higher tax rate and more tax revenue to sustain their costly operation and intra-system higher transaction fees. In addition to the tax policy, are there any other irrational regulations? The answer is definite affirmative, such as cumbersome expense reimbursement system and acquisitive professional title assessment system which, to a certain degree, force clients to choose a bad mode to participate in collusion.

There is a profoundly implicative commentary sentence spread widely among Chinese online communities: "Stringent legislation without taking real circumstances into consideration results in widespread violation of laws and eventually in selective enforcement of laws". This profound remark helps us to understand the institutional environment where collusion is formed. In a society where a qualified enterprise strictly observing laws and regulations finds itself on the verge of bankruptcy, it is ten to one that violating laws or bypassing regulations for survival are very common among all sectors of industry. Once an enterprise is caught skirting laws or deliberately breaking relevant regulations, some local officials will pull rank to bring discretionary power into full play for their own interests. In other words, the local government can abuse its power to exempt the enterprise from liability for its breach of law if the latter agrees with the former's proposal to form an alliance of collusion. The enterprise has no alternative but to collude with the local government. Once trapped in collusion, the enterprise has to expend much time and money on the consolidation of this alliance. In this process, corruption brews, rent seeking emerges, and accidents lurk.

Thus, from the institutional environment of corporate operation, an enterprise, if able to rely on its own normal mode of operation to earn adequate profits, will be less likely to choose the incentives of a bad mode of production or to collude with the local government. In other words, if expected revenue rewarded with lawful operation is higher, the reservation utility will be higher too, participation in collusion will be more costly and incentive compatible constraints of collusion will be more difficult to satisfy. For the central government, deterring collusion becomes much easier.

9.4 Regional Competition and Collusion

Since Tiebout (1956) fully described his hypothesis in his seminal essay "A Pure Theory of Local Expenditures", decentralization and regional competition have been regarded as major driving forces of economic growth and provision of public goods. Some Chinese scholars (Qian and Weingast 1996; Yang 1998; Zhang 1998) also credited regional competition with China's achievement in economic growth. The logic whereby local governments promote economic growth is assumed to be the observation that local governments will make endeavors to improve investment environment, provide better public goods and attract more enterprises and residents voting with their feet.

However, the logic above cannot explain a lot of negative externalities of regional competition such as local protectionism, overlapping projects, environment pollution, tax frauds, etc. Similarly, by that logic we cannot explain the widespread collusion—including not only environment pollution and tax fraud but also coalmine accidents, food safety, and smuggling. It is obvious that regional competition is not only credited with positive effects on the one hand but also it is blamed for negative effects. Nie and Dang (2013) discovered that whichever effect dominates over the other depends on specific conditions.

As far as positive effect is concerned, regional competition seems to be helpful in reducing collusion. Suppose there are a central government, two local governments and an enterprise. When the two local governments (of different regions) compete for the same enterprise, Bertrand (price) competition will cause the local governments to leave all profits to a monopolistic enterprise. At this time, the monopolistic enterprise may obtain much more profits than it does in the case of collusion, so that enterprise under discussion here tends to choose a region where a good mode of production is adopted.

Now let's assume that there are two enterprises E_1 and E_2 which respectively belong in two different regions R_1 and R_2. The enterprises E_1 and E_2 are heterogeneous and E_1 has a more costly mode of production than E_2. On the condition that the two enterprises do not move across the two regions, the local government in R_1 has to collude with the enterprise E_1 and adopts a bad mode of production so as to win over its opponent in the horizontally intergovernmental competition. Simultaneously, the local government in R_2 where the mode of production is less costly predicts the collusion in R_1 and has to collude with E_2 for fear that it should lose its initial competitive edge. Eventually, collusion evolves into a Nash equilibrium of the prisoner's dilemma.

To conclude, regional competition cannot guarantee the elimination of collusion. Whether collusion can be eliminated or not actually depends on specific institutional environment.

9.5 Summary

This chapter mainly discusses the institutional environment of collusion. We hold the view that the central government, under the Chinese-style centralization of authority, cannot take care of every single affair in a hands-on manner and therefore delegate its authority to local governments or some departments directly affiliated to it. However, local governments are not responsible for the local electorate, so information asymmetry will naturally lead to the motivation for collusion. We have explained why the central government or a department directly under the central government periodically deters collusion: in the case of collision, the advantages of economic growth, and costs of maintaining social stability differ in different regions and different periods of time. We have also discovered that stringent legislation without taking real circumstances into consideration coupled with discretionary power in the hands of local governments forces enterprises into the snare of collusion. Regional competition cannot guarantee the elimination of collusion and this depends on specific institutional environment.

The policy implications of this chapter are as follows:

First, for thorough elimination of collusion, the central government should strengthen local residents' supervision over local governments and enterprises, reform the political system, delegate more authority to residents and function by the mandate of the people.

Secondly, the central government's cyclical collusion-proof measures may disrupt enterprises' investment expectation and thus count against the protection of their property right and the improvement of safety facilities.

Third, the central government should readjust those unreasonable and impractical laws and regulations, guarantee the survival and development of those qualified enterprises which lawfully conduct their business in the market, inflict heavier punishment on collusion and instruct, and guide enterprises to break through the snare of collusion.

Fourth, the central government should issue proper laws and regulations, guiding regional competition into a sound development track.

Notes

1. Centralization and decentralization here are discussed in the dimension of administrative system and have no direct relationship with political system. Here we focus on administration efficiency.
2. This can be explained with the following evidence. Under the tax-sharing system, business income taxes should be vested in local governments. However, according to the 2002 *Income Tax Sharing Program*, the central government adjusted the sharing ratio between the central government and local governments to 50:50. But 1 year later, the ratio was readjusted to 60:40 (Lü and Nie 2014). Obviously, the central government can exercise its political authority to unilaterally adjust taxation system.
3. It should be noted that the two concepts of centralization and decentralization are always relative. On the basis of different definitions of centralization, Nie and Zhang (2015) compare the level of well-being of collusion under territorial management (decentralization) and vertical management (centralization).

References

Acemoglu, Daron, Simon Johnson and James A. Robinson. "Institutions as the Fundamental Cause of Long-Run Growth". In Philippe Aghion and Steven Durlauf eds. *Handbook of Economic Growth*. Amsterdam: Elsevier, 2005, vol. 1. 385–472.

Laffont, Jean-Jacques and David Martimort. "Collusion and Delegation". *Rand Journal of Economics*. 1995, 29 (2): 280–305.

Lü, Bingyang and Huihua Nie. Flexible Sharing Contracts: The Contract and Economic Performance of China's Tax-sharing System. *Economic Theory and Business Management*. 2014 (7): 43–50. 吕冰洋, 聂辉华.弹性分成:分税制的契约与影响.经济理论与经济管理. 2014 (7): 43–50.

Nie, Huihua. "A New Perspective of Reducing Safety Accidents". *Globe*. 2005 (1): 11–12. 聂辉华. 减少生产安全事故要有新思维. 环球. 2005 (1): 11–12.

Nie, Huihua and Li Dang. "Can Competition Weaken Collusion?". Renmin University of China, School of Economics working paper, 2013. 聂辉华, 党力. 地区竞争会削弱政企合谋吗? 中国人民大学经济学院工作论文. 2013.

Nie, Huihua, Minjie Jiang and Xianghong Wang. "The Impact of Political Cycle: Evidence form Coalmine Accidents in China". Renmin University of China, School of Economics working paper, 2012.

Nie, Huihua and Yuxiao Zhang. Decentralization, Centralization, and Collusion between Local Government and Central Government. *Journal of World Economy*. 2015 (6): 3–21. 聂辉华, 张雨潇, 《分权, 集权与政企合谋》,《世界经济》, 2015 (6): 3–21.

Qian, Yingyi and Barry R. Weingast. "China's Transition to Markets: Market-Preserving Federalism, Chinese Style". *Journal of Policy Reform*. 1996, 1 (2): 149–185.

Qin, Xiao. "Turning Institutions and Labor Productivity Dividend". *China Reform (Bimonthly)*. 转向制度和劳动生产率红利. 中国改革. 2011 (1): 20–23.

Tiebout, Charles. "A Pure Theory of Local Expenditures". *Journal of Political Economy*. 1956, 64 (5): 416–424.

Tirole, Jean. "Collusion and Theory of Organizations", in Jean-Jacques Laffont, eds. *Advances in Economic Theory: Proceedings of the Sixth World Congress of the Econometric Society*. Cambridge: Cambridge University Press. 1992.

Yang, Ruilong. "The Three Stages of Transformation of China Institutional Evolutionary Mode". *Economic Research Journal*, 1998, (1): 3–10. 杨瑞龙. 我国制度变迁方式转换三阶段论. 经济研究. 1988, (1): 3–10.

Zhang, Weiying and Shuhe Li. "Regional Competitions and Privatization of State-Run Enterprises". *Economic Research Journal*. 1998, (12): 13–22. 张维迎, 栗树和. 地区间竞争与中国国有企业的民营化. 经济研究. 1998, (12): 13–22.

CHAPTER 10

Reform Prospects of the China Model

As the concluding part of the book, this chapter echoes with the initial chapter and discusses the prospects and outlets for the China model. Section 10.1 insists that the China Model is a synthesis motivated by collusion, benefited by demographic dividend, navigated by export substitution, and expressed by extensive economic growth. Section 10.2 holds that China's economic growth is in a special equilibrium of whipping top, which keeps its balance only through high-speed rotation around its own axis and that China should get rid of the equilibrium and achieve a transition from collusion to cooperation. Section 10.3 sums up the weak points of the research and the domains of future research on this topic.

10.1 Demographic Dividend, Extensive Mode of Growth and Collusion[1]

In this section, we will first analyze the reasons for China's extensive economic growth model—resource endowments (demographic dividends), domestic environment (reform route and collusion), and international economic climate (globalization)—and point out the long-term cost of this growth model. Then we will prove by a simple theoretical framework that China may not be able to gain benefits from international trade even if it enjoys the advantage of demographic dividend in an environment where the institutional quality is inferior and transaction fees are comparatively high. In the end, we provide the policy choice for

H. Nie, *Collusion, Local Governments and Development in China*,
DOI 10.1007/978-981-10-5059-6_10

the China model to get out of the mire of extensive economic growth: only by deepening the reform of political and economic systems and improving institutional quality can China sustain its long-term economic growth.

10.1.1 The China Model: Polyhedral Equilibrium

Economic growth is an eternal question that any government should address. The types of economic growth models are usually extensive and intensive. The extensive one is mainly based on the expansion of quantity of inputs to increase the quantity of outputs, while the intensive one is based on technological improvement and innovation to materialize the same objective. There is a consensus that extensive growth is undesirable in the long run because it wastes a lot of resources with an extremely low input–output ratio and that sustainable growth should be based on the intensive growth model. As a matter of fact, the Chinese central government predicted the necessity of transforming the type of economic growth model and set it as one of the objectives of economic reform in the 1980s. The Report of the Thirteenth National Congress of the CPC specified the strategy for economic development was to "gradually shift the emphasis from extensive management to intensive management".[2] Although the Chinese government has set such a strategic objective, economic growth is still mainly based on an extensive model driven by collusion, demographic dividend, and export substitution. This is also the fundamental characteristic of the China model reflected in terms of economic growth.

Granted, in the early stage of reform, it was institutional innovation that boosted economic growth. In the countryside, the family contract responsibility system widely spread throughout the country in 1982 increased the output of foodstuff and agricultural productivity. Almost simultaneously township industries with rural collective and individual economic organizations proliferated, small commodities market and medium- and small-sized manufacturing factories boomed. In cities, state-run enterprises commenced the reform of contract system and joint-stock system; Shenzhen, Zhuhai, Xiamen, Shantou, and Hainan Island were designated as Special Economic Zones (SEZs) in succession; 14 sea port cities including Dalian, Qinhuangdao, and Tianjin were designated as costal open cities. Thus, an open economic landscape radiating hinterland China from costal China came into existence. However,

China's reform seemed to get stalled in the bottleneck where both economic and political reforms entered the deepwater zone and institutional innovation slowed. It is very difficult for us, when looking back on the past two decades, to find an institutional innovation (except for a few cases such as China's entry into the World Trade Organization) that can be as innovative as contract responsibility system and the introduction of Special Economic Zones. The reason for the lack of momentum can be ascribed to the guideline of "incremental reform preceding storage reform", i.e., reforming what is easy before reforming what is difficult. Hence, over time, the cost of reform is additive and accumulative, and naturally, slackens the pace of reform. Furthermore, the yields of old institutional innovation reflect some technical characteristic of the law of diminishing marginal returns.

The above intuitive observations are in consistent with the research results by economic experts. By analyzing China's total factor of productivity (TFP), they have discovered that the TFP growth in the post-1990s slowed when compared with that in the 1980s (Xie et al. 2002; Huang and Ren 2002). This reveals that China's economic growth in the 1990s relied more on the input of labor and capital than on technological innovation. Dong and Wang (2010) used the 1980–1991 macroeconomic data and computed the contribution rates of technological innovation, capital and labor to economic growth to be 0.296, 0.561, and 0.143 respectively, but in the 1990s they were respectively 0.084, 0.771, and 0.145. It is obvious that the contribution of technological innovation to economic growth is significantly decreasing, while the contribution of material input is in the opposite direction and the characteristics of extensive model are more obvious. Both capital and labor inputs are highly associated with demographic dividend.

We have found that, at least in terms of economic growth models, the central government's established policies and the local governments' actual policies are conflicting in the process of achieving economic growth goals. This section tries to address the following two questions: how to understand the conflict between the extensive and intensive economic growth models and how to get out of the trouble?

Of all the factors driving China's long-term economic growth, the contribution rate of population is indispensible. Among all countries of the world, China boasts the largest population, approximately 1370 million, a source endowment that propels the quick economic growth of China. According to the UN's estimates, the proportion of the working-age population (age between 15 and 64) began to climb in the mid-1970s and will peak at 998 million in 2015 (United Nations 2009). Massive working-age population not only relieves the burden of dependency ratio for the whole society but also creates abundant workforce for economic development, thus making it a bonanza known as demographic dividend. Cai and Wang (2005) and Cai (2010) use dependency ratio as the proxy indicator of demographic dividend and found that the contribution rate of demographic dividend to GDP between 1982 and 2000 averaged 26.8%. Since the reform and opening up, demographic dividend has contributed too much to economic growth. In the 1980s when township industry was allowed to enter the market as an important part of private economy, family contract responsibility system released a lot of surplus rural workforce which quickly swarmed into township enterprises and promoted the development of nonagricultural industry in the countryside. In the 1990s when marketization became the economic engine, migrant workers flooded into cities and accelerated urbanization and the development of small- and medium-sized private enterprises. This vividly mirrors Lewis's dual economic theory (Lewis 1954).

The role of demographic dividend is reflected in its shaping of economic growth model over the past three decades. In international trade, surplus workforce signifies comparative and competitive advantages of labor-intensive industry. For China, these advantages, coupled with the comparatively weak domestic consumption market due to low per capita income, contributed to China's prudent choice of development strategy of export substitution. Abundant workforce is equated with low labor cost and high savings rate, which implies that the government-controlled capital is low-cost. The factors, motivated by the Forging Ahead Strategy—bridging the economic gap between China and advanced countries—contributed to the extensive economic growth model characterized with export substitution: launching new projects in different regions and expanding the scale of investment to develop economy. The institutional driving force of extensive economic growth model is collusion. Local officials (representing local governments) are destined to collude with enterprises when they choose low-cost, low-tech, and low-added-value extensive economic growth model to achieve quicker economic growth, more GDP and fiscal revenue, and higher political position. Comparatively, high tech, high-added-value, and environment-friendly intensive economic growth model is almost impossible to achieve instant results in the short term and therefore will be trounced by its opponent (extensive model) in the horizontal regional competition. Such being the case, local governments are reluctant to choose intensive economic growth model, even though they sometimes deny it. In my viewpoint, demographic dividend, export substitution, collusion, and extensive economic growth model constitute a quaternity of polyhedral equilibrium. Although it was proposed in the Party's report as early as in October 1987 to shift economic growth model, this shift has not been yet materialized.

10.1.2 Unsustainability of Extensive Growth Model Driven by Demographic Dividend

The extensive growth model driven by demographic dividend is unsustainable.

First, demographic dividend is going to be gone and Lewis turning point is around the corner. According to the UN, total working-age population is estimated to decline in 2015, and the number of people aged 65 and above will continue to grow until 2030 (United Nations 2009). A society where the aged population accounts for 7% of the total is often defined as an aging society. According to the statistics of the China National Statistical Bureau, China entered an aging society in 2000.[3] Since then, the percentage of aged people has been climbing and reached 8.5% in 2009 and will continue to increase. With the decline of the absolute number of working people, the relative number is declining too. The exemption of agricultural tax, together with the promotion of rural medical insurance and pension insurance, the increase of urban living cost and the rise of opportunity cost of working in cities have resulted in "the shortage of migrant workers" over the recent years (Qin 2011). On the whole, the growth of labor supply is slower than that of labor demand, payroll costs rise, surplus rural labor reaches a financial zero, and urban wages rise drastically (Cai 2010). A research from the Chinese Academy of Social Sciences reveals that the marginal contribution of the increase of the general quantity of workforce resource to economic growth is diminishing and that China cannot count on demographic dividend forever (Research Group of Economic Growth and Macroeconomic Stability in China 2007).

Second, a limited quantity of energy resources is hard to sustain quick growth. In the case of the extensive model, quick economic growth requires a continuous supply of energy resources. Table 10.1 shows

Table 10.1 Energy consumption of major economies

Economies	*Energy production (MTOE)*	*Energy consumption (MTOE)*	*Energy consumption growth rate (%)*	*Net imports/consumption*
	2008	*2008*	*1990–2008*	*2008*
China	1993.3	2116.4	5.1	5.8
Japan	88.7	495.8	0.6	82.1
USA	1706.1	2283.7	0.95	25.3
World	12,356.8	11,897.3	1.8	
China's rank	1	2		
China/World	16%	18%		

Note MTOE refers to million tons of oil equivalent
Source World Bank Development Indicator (http://data.worldbank.org/indicator)

that China, as the largest developing country in the world, is the largest energy producer and the second largest energy consumer. From 1990 to 2008, the average growth rate of energy consumption increased drastically, surpassing the average level of major advanced countries and all other countries of the world. Compared with 1990, China has shifted from an energy exporter to a net energy importer. Extensive growth model resulted in a massive shortage of electricity in the southern part of China for many years. It is reported that in the spring of 2011 provinces are short of electricity supply.[4] Within the same period, China also consumed a third or a half of the global total consumption of key resources such as iron ores, coal, lead, zinc, aluminum, copper, and nickel.[5]

Third, long-term extensive growth model causes serious environment pollution. Compared with intensive model, extensive model features higher input, lower output, and heavier pollution. According to experts' estimates, in the 1990s, about 4–6% of the annual GDP growth rate of 9.8% was achieved at the cost of environment degeneration, including natural resource depletion, ecological disruption, and pollution.[6] Table 10.2 reveals that China is one of the largest energy consumers and also one of the least efficient energy users. The production of GDP per unit is lower than that of advanced countries and even lower than the global average, ranking nearly at the bottom. From the result of energy consumption, the total emission of major pollutants ranks first. Due to the obvious negative externalities, environmental pollution resulting from extensive model causes the resentment of neighboring countries.

Table 10.2 Results of energy use of major economies

Economies	*GDP per unit*	*Emission of carbon dioxide*	*Emission of methane*	*Emission of carbon oxide*
	2008	*2007*	*2005*	*2005*
China	3.6	6533	955,760	566,680
Japan	8.1	1253.5	53,480	23,590
USA	5.8	5832.2	810,280	156,210
World	5.5	30,649.4	6607,490	3787,800
China's ranking	104	1	1	1
China/World		21%	15%	15%

Notes (1) The source is from World Bank Development Indicator (http://data.worldbank.org/indicator); (2) The unit of energy is per kilo of oil equivalent, GDP is measured in terms of PPP (power parity) dollar in 2005; the unit of carbon dioxide is million tons; the unit of methane and carbon oxide is thousand tons of carbon dioxide equivalent

Fourth, excessive export substitution has resulted in trade frictions. Cheap labor makes its manufactured and assembled goods enjoy substantial cost advantages, which is one of the major reasons for industrialized countries to transfer their manufacturing to developing countries including China and also for other emerging economies to grow rapidly over the past decades. However, the flood of China-made goods has had a great impact on low-end markets and employment of the importing countries, thus worsening the trade frictions between China and some advanced countries. As of June 30, 2010, China has received 119 anti-dumping complaints, next only to the USA (257) and the European Union (149) but high above Canada (36), South Korea (37), and Japan (6).[7] Frequent occurrence of trade frictions is an unfavorable condition for China to peacefully rise on the world stage.

Nevertheless, demographic dividend will not necessarily result in energy over-depletion, environment pollution, and frequent trade disputes. The above problems worsen when demographic dividend is combined with export substitution and extensive growth model and motivated by Forging Ahead Strategy. India, a China's neighbor also with a large population (1200 million by 2011), has not suffered from serious consequences because it chooses a service-motivated strategy over industry-motivated one and its government cannot promote a forging ahead strategy like China under the democratic political system and federalist national system. As a matter of fact, India's quick growth over the past 10 years relies on its innovation of banking system and privatization (Huang 2012).

Comparatively, India's growth model is more intensive than China's even though it currently lags behind China in terms of economic growth.

10.1.3 *A Theoretical Analysis: Transaction Cost Neutralizing Demographic Dividend*

In this section, we are going to explain why China, even if its demographic dividend could sustain, may reap less revenue in the international trade than what the traditional trade may expect and why the social well-being bestowed by international trade, under a certain condition, might be lower than that not bestowed by international trade. Let's describe a standard Heckscher-Ohlin trade model. Suppose there are two countries—Country North and Country South, two factors of production—capital and labor, two products—capital-intensive product and labor-intensive product. Country North is comparatively abundant in capital while Country South in workforce. Then, according to comparative advantage theory, Country North should produce the capital-intensive product and Country South should produce the labor-intensive product. When the marginal return of capital in Country North is equal to that of labor in Country South, they mutually exchange their products. The benefit of international trade is the improvement of well-being in both countries. But the traditional trade theory faces two universal paradoxes: first, the advanced country abundant in capital does more trade intercourses with other advanced countries than with less-advanced countries abundant in labor, not theoretically the other way around; second, international trade worsens (not theoretically improves) the intra-national and international income inequality (Wood 1997; Kremer and Maskin 2006). We argue that the traditional trade theory cannot explain the above paradoxes in that it ignores institutional quality. Taking this into consideration, we attempt to construct a temporary theory containing institutional quality for our purpose: to explain the two paradoxes and to discuss the prospects of the China economic model characterized with demographic dividend.

Now we add an extra condition—institutional quality—to the supposition in the previous paragraph. Suppose the advanced Country North enjoys a superior institutional quality and lower transaction cost while the less-advanced Country South has an inferior institutional quality and higher transaction cost.[8] Meanwhile, we also suppose that the consumers in both countries can move across borders to some degree (not migrants). According to the theory of comparative advantage, Country North should produce the capital-intensive product and Country South should produce the labor-intensive product and mutually exchange products. When the institutional quality of South Country is inferior, the domestic transaction cost[9] of the product exceeds its export cost to the extent that the price of the Country South's exporting product in domestic market is higher than in Country North's market but the domestic profit is lower than overseas profit. Conversely, due to superior institutional quality and lower transaction cost, Country North sells its product at a much higher price in overseas markets than in domestic markets. In this case, consumers in Country South flow into Country North to purchase both the capital-intensive product that also should be exported to Country South and the labor-intensive product that is imported from Country South. Economies of scale and competition effect result in the low price of both labor-intensive and capital-intensive products. The market in Country South gradually shrinks because the selling prices of both products are higher. In a most extreme condition, both of the products are sold only in the market of Country North where its capital-intensive product can only gain exporting revenue without being exported to Country South. So long as the prices of the two products in Country North are lower than those in Country South, Country North will be the monopolist of the two products.

What will be the consequences of the trade between the two countries? From the perspective of the well-being of the consumers, part of the surplus of the consumers in Country South is squeezed out by consumers in Country North and therefore the purchasing power of the consumers in Country South will be weakened with the expansion of the market in Country North. From the perspective of the well-being of the manufacturers, the profits that the manufacturer in Country South can earn is dwindling due to the shrinkage of domestic market arising from the fact that a large number of consumers in Country South go to Country North and purchase both capital-intensive and labor-intensive products. Owing to simultaneous dwindling of the surpluses of the consumers and manufacturers in Country South, its well-being drops on the whole; conversely, social well-being of Country North improves with the increase of the number of consumers. In the most extreme condition, much more loss of well-being may result in an absolute drop of its well-being even in the case of its trade relations with Country North. In the long run, due to the drop of consumers' purchasing power and a decrease of producers' profit, Country South has not enough capital to make technological innovation, thus unable to dynamically update its industrial structure but to preserve its comparative advantage of cheap labor. Eventually Country South gets into the trap of low-end division of labor.

The transaction-cost-containing international trade theory proposed here is different from traditional trade theory. Some new conclusions are reached on the basis of combining supply and demand. First, when the transaction cost in the less-advanced country remains high, there will be fewer trade opportunities between the two countries. The less-advanced country serves as production bases and its processed products will be exported to the advanced country which, however, exports much fewer products to its counterpart. Second, international trade is likely to widen the well-being gap between the two countries, thus worsening the income inequality. Third, the less-advanced country counting on the comparative advantage of labor is likely to be locked-in the low level of the chain of labor division in the long term. Obviously, inferior institutional quality or higher transaction cost will erode the advantage of demographic dividend and that the economy based on this advantage is unsustainable.

Our theory can be expanded to a situation where many other countries are included. Given many countries in the world, there is the most extreme condition in which a labor-intensive country with the largest population and the highest transaction cost manufactures products for all other countries ("workshop of the world") and its residents go to other countries to purchase products. As a result, this country will face the following problems such as low profit margin and consumptive level, high energy consumption and serious environment pollution while other countries may enjoy the fruit of "institutional dividend". In this sense, the epithet of "workshop of the world" is not the metaphor of high profit, but of wage slaves.

Some cases in the real world testify to our theory. Comparatively, due to the imperfection of market economy system and of legal system, mainland China's taxes, gray costs, public relation fees, and the cost of property rights protection are higher than those in overseas markets, thus making the transaction cost in mainland China much higher than in overseas markets. This is the reason why the China-made products are cheaper in overseas markets than in its domestic markets (Nie 2010). For instance, Nike athletic shoes are sold at CNY 700 to 800 per pair in

China, but at only CNY 200 or CNY 300 (converted in terms of USD/CNY exchange rates) in the USA; a suit of Armani men's garments is sold at as high as CNY 30,000 in China, but at only CNY 7000 in the USA.[10] Similar cases are numerous.[11] It is estimated that, due to overseas (online) purchasing for which domestic high transaction cost should be blamed, domestic manufacturers suffered a heavy loss in sales volume worth at least CNY 12,000 million (Wu 2011). From the perspective of producer surplus, domestic producers earn extremely low profits as OEMs (original equipment manufacturers) or contracting manufacturer for foreign enterprises. Take the currently popular iPhone 4, Chinese producers can get only seven US dollars per unit of the aggregate net profit of 360 US dollars.[12] In addition, the exact loss of consumer surplus and producer surplus arising from domestic high transaction cost is hard to divine. This is no doubt a question worth a research.

10.1.4 Way Out: Improving Institutional Quality Through Reforms

Traditional international trade theory holds that it is the degree of abundance in factors of production that determines comparative advantage theory which in turn determines division of labor and pattern of trade. Suppose capital is rarer compared with labor and thus capital-intensive products are more tech-specific and profitable. Then, according to traditional trade theory, a country should be dedicated to develop labor-intensive industry and acquire advanced technology so as to accumulate more capital for subsequent updating of industrial structure from labor-intensive model to capital-intensive model. This is so-called dynamic comparative advantage theory.

Our point of view is that a country with superior institutional quality is competent enough to decrease transaction cost and production cost, and thus obtain more consumer surplus and producer surplus. In the real world, due to imperfect contract, specific investment faces the opportunistic behavior of holdup. In a country with superior institutional quality, its contracts will be more complete and the level of specific investment will be accordingly higher (Nunn 2007). According to Nunn's definition, we have found that industrial sectors with higher contract density (such as transportation equipment, telecommunications equipment, general-purpose equipment, and instruments) are usually capital-intensive industries. When capital is rarer relative to labor, the average profit margin of capital-intensive industries is higher than that of labor-intensive industry, an observation from which we can roughly deduce that contract density is in positive correlation with profit margin. This denotes that a country with superior institutional quality is more likely to obtain higher profits in the international division of labor. It is by no means easy to testify to this conclusion from an econometrical approach, but in economics literature (such as Acemoglu et al. 2001), the idea that institutional quality is in positive correlation with per capita income seems to be, at least temporarily, indisputable. According to our newly modified trade theory, a country cannot hold an advantageous position in the international division of labor unless it has improved its institutional quality (Nunn 2007; Levchenko 2007) and make full use of institutional dividend.

Granted, according to our previous analysis, demographic dividend can help a country or region obtain cost advantage in the international division of labor and thus quickly bring about economic growth by relying on export substitution or extensive growth model, but this growth model will bring tremendous cost to both China and the rest of the world in the long run and thus it is unsustainable. As a fundamental factor for economic growth, institutional quality has the characteristics of sustainability and political consolidation (Acemoglu 2005) and therefore is unlikely to improve within a short period of time. Theoretically, demographic dividend and institutional dividend are not necessarily replaceable but mutually complementary. The core is what development strategy a country should choose.

For our purpose, we might as well, on the bi-dimensional basis of institutional quality and demographic dividend, divide all nations or regions into four types (See Fig. 10.1): (1) lower demographic dividend and inferior institutional quality (quadrant III); (2) lower demographic dividend and superior institutional quality (quadrant II); (3) higher demographic dividend and inferior institutional quality (quadrant IV); (4) higher demographic dividend and superior institutional quality (quadrant I).

We use total dependency ratio to measure demographic dividend, an age-population ratio of those typically not in the labor force or the dependent part aged below 15 and above 65 to those typically in the labor force or the productive part aged between 20 and 64. A lower numerical value of the ratio implies a higher demographic dividend. The relevant data are from the UN's demographic data (United Nations 2010). Then we use ease of doing business index to measure institutional quality, a measure of regulations directly affecting businesses

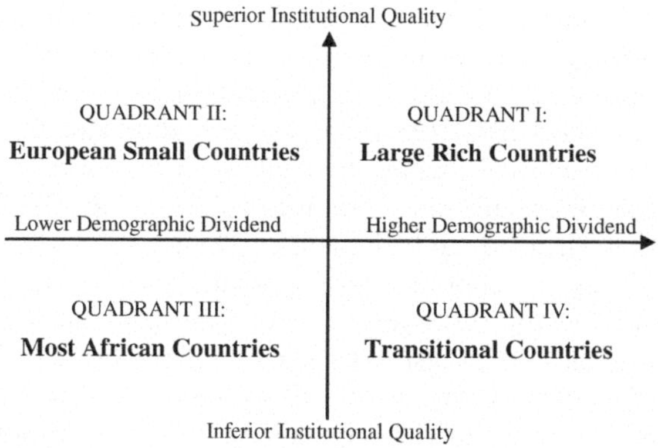

Fig. 10.1 Demographic dividend and institutional quality

which include the following subindices: starting a business, dealing with construction permits, registering property, getting credit, protecting investors, paying taxes, trading across borders, enforcing contracts, and resolving insolvency. A lower numerical value (or a higher ranking) of ease of doing business ratio indicates superior institutional quality. The relevant data of this index are from World Bank's *Doing Business Report 2011*.[13]

Different types of countries have their own distinctive competitive profiles and choose different development strategies. The first type (Quadrant III), due to lack of the comparative advantage of labor in the short term and of institution in the long term, can but passively get involved in international division of labor and will be in a state of long-term stagnancy unless dramatically reverse changes take place. Based on total dependency ratio and ease of doing business ratio, this type of country includes Afghanistan, Nepal, Angola, Democratic Republic of Congo, and Chad etc. For these countries, the top priority is to solve the problem of survival: to quell unrest or war, prevent the spread of diseases, prolong life expectancy, multiply working population, and gradually improve institutional quality in the long run. The second type (Quadrant II) lacks the comparative advantage of labor but enjoys superior institutional quality. This type includes France, Switzerland, Colombia, Israel, Peru, and many other European and South American countries. These countries, due to superior institutional quality and higher per capita income but lack of working-age population and its concomitant disadvantage in labor-intensive industry, depends more on its trade relations with less-advanced countries. These countries should consolidate their established institutional advantages and optimize population structure by encouraging fertility. The third type (Quadrant IV) has abundant workforce but lacks superior institutional quality. It includes those countries which are undergoing economic and social transitional, such as Russia, Poland, Romania, China, Thailand, and Vietnam etc. Due to abundant workforce despite delayed start in economic development, they can still gain quicker economic growth in the short term on the one hand but comparatively inferior institutional quality restricts their development in the long term on the other. Countries falling within this category will be promoted to the club of rich countries in the event of successful institutional reform; otherwise, they will still lumber in the long procession of middle-income countries, that is, they will fall into the middle-income trap. For these countries, the top priority is to set their sights on improving institutional quality and reducing transaction cost. The fourth type (Quadrant I) is the most ideal club of rich countries, which boast abundance workforce and superior institutional quality and therefore are very competitive. This club includes Singapore, South Korea, Canada, Sweden, Australia, Germany, the UK and the USA, etc. It is notable that both Singapore and South Korea have successfully got out of middle-income trap and completed the transformation from demographic dividend to institutional dividend.

China belongs to the third type. Its total dependency ratio ranks the 18th place (higher demographic dividend) and its ease of doing business index ranks the 79th (inferior institutional quality). Currently, China's per capita GDP is more than \$3000 and has entered into middle-income group with per capita GDP ranging between \$3000 and \$5000. It is now facing the challenge of middle-income trap.[14] According to our previous analysis, China's demographic dividend can only sustain a short period of time and cannot continue to be the important momentum for economic growth. It is notable that some emerging Asian countries like Thailand and Vietnam, owing to their still much cheaper labor, have become the reserve base to which advanced countries transfer manufacturing industry. In addition, economic and technological powers represented by the USA still control sophisticated technologies and key industries and their leading edge is impregnable at least in the short term. Therefore, in the long run, China must improve its institutional quality so as to occupy a favorable position in the international division of labor and upgrade itself from Quadrant IV to Quadrant I, obtaining institutional dividend while continuing to preserve its advantage in population scale or comparative demographic dividend.[15] Otherwise, China would lose not only its advantage of labor cost but also its competency in some high-tech and key sectors of industry.

For China, the specific approaches to superior institutional quality include: cutting off the chain of collusion through legislation, protecting the public's properties and rights, protecting intellectual property right and technological innovation, curbing corruption, improving the government's administrative efficiency, quickening the development of legal system and protecting contractual relationship. China has led developing countries and even some developed countries in the construction of infrastructure, but it still lags behind many other countries in improving soft power. For the above goals to be attained, there is no alternative but to deepen political and economic reforms. On the one hand, China should make full use of demographic dividend that is still sustainable within the next few years, accumulate capital and improve technological level and make preparations for forthcoming dynamic industrial upgrading. On the other hand, China can combine short-term demographic dividend with long-term institutional construction, for institutional construction entails labor input. Only when China takes advantage of its current demographic dividend to realize its long-term institutional dividend, can it eventually be promoted to a leading power in the club of rich countries.

10.2 Out of Whipping Top Equilibrium: From Collusion to Cooperation

10.2.1 Trap of Whipping Top Equilibrium

Between 1980 and 2010, China was the fastest growing economy and its GDP increased at 10% per annum. Why did it need such a fast economic growth rate? What if a slower rate? How long can such a quick growth rate sustain? At present, the theoretical community has not provided a satisfactory answer.

We argue that, given the current political and economic systems and international environment, China could not decelerate its pace within a short period of time. Once its economy slows down, a lot of economic, social and even political problems will follow on the heels of it. As a matter of fact, China's economy landed into a whipping top equilibrium, featuring fast growth rate, tremendous input, high energy consumption, and heavy pollution versus low-end technology, inferior quality, low price, and low profit. All these inseparably interconnected features reflect the multifaceted characters of the China model. Whipping top equilibrium can be simply defined as "high economic growth concomitant with frequent accidents". A whipping top is a toy that could spin rapidly around its own axis and remain precisely balanced on its tip because of its rotational inertia exerted by a whip; without continuous force exerted onto it, it will stop and fall. China's economy is just like a whipping top, remaining balance only when growing fast. The forces that propel China's economy are institutional force (collusion), workforce (demographic dividend), and pulling force (export substitution).

At present, the reasons that China must maintain quick economic growth are as follows. First, the ruling party needs to consolidate its governing legitimacy through economic achievements. Second, only through economic growth, can the mass of population could have the opportunity to be employed. Third, serious social problems such as corruption, pollution, and waste devour a considerable amount of economic achievements. Fourth, the global economic chain has adapted itself to China's rapid economic growth.

However, like a whipping top which is by no means a perpetual-motion machine, China's spinning economy could not receive a nonstop driving force. As is mentioned above, China's economy based on extensive economic growth model is unsustainable due to the hard constraint of shortage of resources. Until now, throughout the world history, no country could maintain its economic growth rate at 10% per annum for as long as half a century. Therefore, the heart of China's economic problem is to slow it down and maintain a moderate growth rate mainly focusing on the improvement of institutional quality and upgrading of industrial structure. The USA set a good exemplar. As No. 1 economic superpower, the USA has always maintained a steadily slow rate over the 100 years (except a few years) between 1908 and 2008. In view of its huge economic aggregate, a steadily slow rate also denotes an impressive increase of the absolute value of economy.

How will China be able to guarantee small quick steps on the trot in economic development for 100 years as the USA has done? The governments at different levels should set their sights on the following commitments.

First, cut off the chain of collusion through strict legislation. For a long period of time in the future, it is the government (including the central government and local governments) but not enterprises or the market that will continue to be the leading force to drive China's economy, so the government's incentive is of vital importance. If local governments lack foresight and still attach great importance to short-term economic results, growth acceleration, and GDP race, they will continue to collude with enterprises and the whole economy will continue to depend on high growth rate, thus being unable to get out of the trap of whipping top equilibrium. Cutting off the chain of collusion entails the reform of political system, especially of performance assessment system for government officials, including: (i) designing a set of effective evaluation criteria or indicators that can reflect the quality of development, such as environment protection, energy consumption, education, and public services; (ii) in assessment methods, ramping up local residents' assessment of local, officials and strengthening the accountability system of local people's congresses for local governments and officials; (iii) in the selection of officials, further pushing grassroots democracy and completing and perfecting the electoral system.

Second, press ahead with the reform of political system so as to consolidate the political legitimacy of the ruling party. The vitality of the Communist Party of China springs from its continuous and unrelenting advance in putting forward new doctrines of governance. Since its seizure of power, the Communist Party of China has always been revising its governing concepts throughout different historical periods. At the beginning of the People's Republic of China, the Mao-led first generation of leadership insisted "taking class struggle as the key link". At the commencement of reform and opening up, the Deng-led second generation of leadership cried out for "focusing on economic construction". In the New Era (in the 1990s) the Jiang-led third generation of leadership proposed the thought of "Three Representatives". In the first decade of the new century, the Hu-led fourth generation of leadership put forward "Scientific Development Concept", which reflects the Party's resolution to shift its sheer reliance on economic growth to top-quality comprehensive development. This lays a solid political foundation for future tuning into a sound and substantial growth model. In the future, China needs to extend democracy to the grassroots level, complete and perfect electoral system so that the ruling party could reinforce its governing legitimacy through economic achievements and political procedures.

Third, encourage enterprises to make technological innovation so that they could entrench their place in the world market by relying more on innovation than on cheap factors of production. Currently, the difficulty that China has in shifting its extensive growth model to intensive growth model lies in the fact that many enterprises lack innovational spirit and core technology and have to compete with international famous brands through labor, land, and profit control. If China has technological advantage, it will naturally give full play to comparative advantage and pursue a technology- and knowledge-intensive growth model. The key to encouraging innovation in enterprises lies in the protection of material and intellectual property rights. The principle of institutional economics shows that the protection of property rights is the foremost important institutional arrangement, an incentive mechanism as which nothing else can be so effective.

Fourth, readjust China's position in the global supply chain step by step through replacing demographic dividend with institutional dividend. Endowed with the largest population and workforce, China should take this advantage to push up institutional construction, gradually adjust its place in the global chain of labor division and evolve into the high-end of the chain. In this way, the global supply chain will gradually adapt itself to these changes and thus trade frictions and tumultuous economic situations can also be substantially reduced.

10.2.2 From Collusion to Cooperation

The four commitments above can neither be fulfilled at one go. Long though it may take to fulfill these commitments, local governments can still get somewhere in the process of abolishing collusion and pursuing intensive economic growth model within a short period of time even under the currently existing system. Read the following case first.

10.2.2.1 The Story of Wenzhou: from forgery to anti-counterfeiting[16]

At the beginning of the 1980s when reform and opening up was still in the initial stage, the Wenzhou model rose to fame and its brisk private economy was the subject which people waxed lyrical over. Shoemaking is such a time-honored craft in this city that as early as in the Ming Dynasty (1368–1644) Wenzhou-made shoes were among the tributes presented to the imperial court, and by the early 1920s, a complete chain of shoemaking had come into being. At the bugle call of reform, savvy

Wenzhou people set up their family leather shoe workshops throughout the city overnight and began their primitive accumulation of capital in the cutthroat competition during which the owners of these workshops do not observe commercial ethics in their business. At that time, China just freed itself out of the period of shortage economy and domestic demand was active but the corresponding laws had not been issued as to the protection of consumer rights. In other words, the institutional environment allowed for forgery. Wenzhou-made shoes were notorious for their inferior quality, standing as a metaphor for similar Wenzhou-branded shoddy and counterfeit goods. For a time, shopping malls or small shops across the country rejected to purchase any products manufactured in Wenzhou, and some of them even hung a board outside reading "No Products Made in Wenzhou in the Mall". Wenzhou shoemakers were facing the risk of reputation crash.

A report reprinted and spread widely was very dramatic. A northeastern young woman bought a pair of Wenzhou shoes for her fiancé. By the third day at the wedding banquet, the bridegroom's shoes wore out. The shoes turned out to be made from strawboard. Seething with anger, the couple sent the worn shoes to the Mayor of Wenzhou and tagged a line to it: "Do you, as Mayor of Wenzhou, feel ashamed of the despicable business conduct in your city with such a rip-off?"

On August 8 (a day which many Chinese merchants believed to be lucky and can bring them good fortune), 1987, in the downtown Wulin Square in Hangzhou, capital of Zhejiang Province, officials from the Bureau of Industrial and Commercial Administration burnt to ashes over 5000 pairs of shoddy shoes made in Wenzhou. The fire, called the "flame of shame", sparked Wenzhou entrepreneurs' awareness of quality and reputation. In June 1988, the first shoe industry association in China was established in Wenzhou. The association appealed to shoemakers to defend the reputation of the shoe capital with first-rate quality. A second shoe-burning event took place on December 15, 1999, when the President of Aokang Footwear burnt over 2000 shoddy shoes.

In 2002, Wenzhou designated August 8 every year as Good Faith Day. On the sixth Good Faith Day, Wenzhou top officials and 52 local entrepreneurs convened in Hangzhou and published Good Faith Declaration: cherishing credit as the life of our business and as the future of Wenzhou. The reputation of Wenzhou Shoes turned around and Wenzhou-made products experienced a journey of rejuvenation. Wenzhou now boasts 58 Chinese Well-Known Trademarks, 32 China

Top Brands, and 34 national production bases, and has risen to one of the most important production bases of advanced manufacturing and top brands.

10.2.2.2 The revelation of Wenzhou: government-firm cooperation guaranteeing win-win

Wenzhou's shift from forgery to anti-counterfeiting reveals that, even if under the current institution, local governments can choose cooperation over collusion with enterprises. The choice of a good mode of production in the shoemaking industry has brought about a virtuous circle of all other sectors of the local industry, boosting economic growth and prosperity.

Wenzhou city government proposed with the National People's Congress an initiative that the city should be upgraded a higher administrative level in 1987 when Wenzhou came to a critical moment after years of initial development based on its advantage of flexible commercial policies, self-employed and private businesses and brisk markets for small commodities. Wenzhou, a city which prospered before many other cities, was confronted with an option: whether to continue a bad mode of production to achieve quick short-term growth or pursuing a more sound and sustainable development road through brand strategy. As a matter of fact, the establishment of a complete chain of shoemaking and the brisk demand for quality products with the improvement of living standards prompted the city government and the shoe industry association of Wenzhou to crack down forgery and stress brand strategy in the hope of earning more expected revenues. Although the central government did not alter the assessment criteria for performances of Wenzhou local governments, the Wenzhou city government voluntarily gave up collusion with shoe makers and chose cooperation; and more importantly, the Wenzhou local government was of great importance in creating the "Wenzhou Economic Model", which inspires other cities to follow suit for modernization.

Unlike collusion, cooperation means that local governments and enterprises choose superior production techniques with top quality, low energy consumption, and environmental friendliness. Since there is no information that could be concealed between local governments and enterprises in the process of cooperation, they can cooperate open and above board and win support from the central government.

However, it would be naive to believe that collusion will automatically cease to exist simply because a local government (such as the Wenzhou city government) voluntarily terminates collusion in a specific industry (shoemaking) under a specific historical situation. What China should rely on is institution, only through which will a successful model be likely to be modeled. Therefore, it should be stressed here that only through institutional reform—such as reforming the current assessment criteria for local officials and election and supervision systems, democratizing grassroots election and perfecting legal system—can cooperation between local governments and enterprises be propelled top-down, thus creating institutional dividend for the likelihood of China's economy occupying a favorable position in the world high-end markets.

10.3 Drawbacks of This Book and Future Research Topics

In the age of high specialization of knowledge, this book attempts to be ambitious enough to expound the China model characterized with high-speed growth and high-frequent pernicious accidents. However, as a Chinese expert in economics, I should confront head-on some issues of great realistic significance and undertake some responsibilities for social development and well-being. Therefore, my attempt, challenge, and innovation are of some values.

This book still has some aspects (listed below) requiring further research although it has built a tentative analytical framework for collusion, explained the problems of work safety and accidents through mathematical modeling and econometric analysis, explored the impact of collusion on enterprises, society and social development, and mentions the institutional source of collusion. I sincerely hope other experts interested in game theory concerning China's central government, local governments and enterprises, be they from economics, politics and law or from management, sociology and history, make a joint effort to apply themselves to an insightful research of this program, exploring a fundamental guideline for national wealth and strength.

1. To analyze the impact of collusion on other industries. This book and other relevant economic papers have analyzed the impact of collusion on coalmine accidents, illegal use of land, housing bubble and tax frauds, but we lack systematic empirical evidence for other issues such as environment pollution, food safety, smuggling, regional externalities, fakery, and forgery. A key problem is

to find an indicator that is easily applied to measuring collusion in a particular field.

2. To continue with research on the impact of collusion on enterprises, governmental behaviors, political performances and social development. Although these issues have been discussed in Chaps. 6–8, theoretical bases and econometric tests associated with them need to be improved. Still, such a problem as how collusion affects governments' policymaking has not been dealt with at all. If following this route, the economics community may enlarge and perfect political economics literature on the making of economic policy. In this respect, the modeling method put forward by Dixit (1996) is perhaps applicable.
3. To delve into the historical origin of collusion. The current situation about collusion has much to do with the historical fact that Chinese enterprises, since at least the Self-Strengthening Movement (1860s–1890s) in the late-Qing dynasty, have been attached to the government. This is quite different from that of Europe. How does the relationship between local governments and enterprises affect the present collusion? Why has an equal contract relation between government and enterprise not been developed?
4. To differentiate and analyze the difference between the core concept of collusion and important concepts such as political connections, corruption, and rent seeking. Compared with other relevant concepts, what new can collusion tell us? What theoretical contribution can it provide with economics literature? How is the relation between corruption and other concepts such as political connections, corruption, and rent seeking clarified theoretically?
5. To analyze collusion from a global perspective. Has there been collusion in Russia and transformational Eastern European countries? Have the Asian advanced countries like Japan and South Korea ever experienced the shift from collusion to cooperation? Does collusion still exist in such authoritarian countries as Vietnam and Singapore? Have some advanced countries like the USA and UK also experienced the stage of collusion?
6. To analyze the impact of collusion on macroeconomic growth. Will collusion result in cyclical fluctuations of macro-economy? How is collusion related with political-business cycle? Do

technological change, exogenous demand or supply shock have direct influence on collusion?

7. To research the relation between collusion and public choice. What is the relation between political and economic systems and the forms of collusion? This book neglects the analysis of the role of local residents. How do they play their roles in fighting and disintegrating collusion? And how do they affect the game playing between local governments and enterprises?
8. To discuss the differences of building coalition of collusion in different types of enterprises. How is different when state-run, collective, private and foreign-funded enterprises get involved themselves in collusion? How does the competition among them affect the formation and disintegration of collusion?
9. To analyze the change of collusion from a dynamic angle. For instance, some enterprises in some regions were dubbed as red cap enterprises (the reputation a private enterprise enjoys by becoming attached to a government department and doing business in the name of a state-run or collective-run enterprise). For a time they became independent but now they are attached to a government department once again. If collusion is dynamic, what are the causes for and effects of this dynamic change?
10. To design the optimal reform strategy under the collusion context. Determined by the Chinese specific conditions, collusion, centralization, growth priority, demographic dividend, and incremental reform path are complimentary to each other, which implies that a substantial reform may trigger a domino effect or a chain reaction. Therefore, the next reform top-level design must take a comprehensive consideration of the above and work out a contingency plan.

I believe, except the above ten, that there may be some other issues which we may neglect. But this does not mean they are of little importance. China is developing fast and problems emerge one after another, and great challenges confront us. It is universally acknowledged that social development and reform is preceded by theoretical innovation, which, I believe will eventually guide China to the gear of rejuvenation.

Notes

1. This section is rewritten and modified on the basis of Nie and Zou (2012).
2. See "Advance Along the Road of Socialism with Chinese Characters", the report delivered by Zhao Ziyang (elected as General Secretary of the CPC in 1987) at the Thirteenth National Congress of the Communist Party of China on October 25, 1987.
3. Chu, Fuling "China Entered into an Aging Society in 2000", http://news.xinhuanet.com/fortune/2012-08/07/c123543329.htm (Accessed September 15, 2012).
4. "A Record of Shortage of Electricity in Eleven Provinces", *The Economic Observer*, April 30, 2011. http://www.eeo.com.cn/2011/0430/200329.html (Accessed May 16, 2011).
5. "Crowded out", *The Economist*, September 24, 2011, http://www.economist.com/node/21528986 (Accessed March 20, 2012).
6. "How to Contain the Increase of Environmental Costs?" *Economic Information Daily*. July 2, 2005.
7. See the WTO document in 2011 (http://www.wto.org/english/res_e/booksp_e/anrep_e/trade_profiles10_e.pdf).
8. Transaction cost here may come from contractual incompleteness (Hart 1995) or from information asymmetry.
9. Transaction cost in this context, in its broad sense, refers to the cost of institutional operation, specifically expressed as the costs of taxes, information, corruption, property rights protection and contractual operation.
10. "Why are China-Made Products More Expensive in Domestic Markets Than in Foreign Markets?" *Nanfang Daily*, November 30, 2010. Column A17.
11. "Lower Wages, Higher Prices, Why?" *China Youth Daily*. June 29, 2011, Column 9.
12. "Warning and Striving: A Reflection on China's Self-Owned Brands". *Economic Information Daily*. November 22, 2010.
13. Since we are mainly concerned about the issue of the environment of international trade, we choose ease of doing business index that measures regulations directly affecting businesses over other institutional indicators (such as Polity IV) that measure political institutions. The data are from *Doing Business Report 2011,* which can be retrieved any time through the website http://www.doingbusiness.org/.
14. Translator's Note: According to the latest World Bank Classification, China falls under the category of upper-middle income countries with per capita GNI (gross national income) ranging between $4126 and $12, 735, calculated using the World Bank Atlas method. See "Country

Income Groups (World Bank Classification)", http://chartsbin.com/view/2438. Retrieved on December 11, 2016.

15. Translator's Note: In October 2015, more than 2 years after the publication of the original book, the Chinese government decided to implement the universal two-child policy, hoping to turn around aging trend.
16. This case is arranged and adapted according to the following news commentary or news interview: "From Forgery to Anti-counterfeiting: Survival Philosophy of Wenzhou Shoemakers", in *Xinmin Weekly*, March 9, 2005; "Lighting the Fire of Commercial Integrity at Wulin Gate in Wenzhou", in *Wenzhou Daily*, August 17, 2007.

References

Acemoglu, Daron, Simon Johnson and James A. Robinson. "The Colonial Origins of Comparative Development: An Empirical Investigation". *American Economic Review*. 2001, 91: 1369–1401.

Acemoglu, Daron, Simon Johnson and James A. Robinson. "Institutions as the Fundamental Cause of Long-Run Growth". In Philippe Aghion and Steven Durlauf eds. *Handbook of Economic Growth*. Amsterdam: Elsevier, 2005, vol. 1, 385–472.

Cai, Fang. "Demographic Transition, Demographic Dividend, and Lewis Turning Point". *Economic Research Review*. 蔡昉. 人口转变、人口红利与刘易斯转折点. 经济研究. 2010 (4): 4–13.

Cai, Fang and Dewen Wang. "Demographic Transition: Implication for Growth", in Garnaut and Song, eds. *The China Boom and its Discontents*, Canberra: Asia Pacific Press. 2005.

Dixit, Avinash K. *The Making of Economic Policy: A Transaction Cost Politics Perspective*. Cambridge, MA: MIT Press, 1996.

Dong, Zhiqing and Linhui Wang. "The Sources of China's Economic Growth: Empirical Evidence from Capital-Embodied Technical Progress". *Jilin University Journal (Social Sciences Edition)*, 2010 (4): 131–138. 董直庆, 王林辉. 我国经济增长来源—来自资本体现式技术进步的经验证据. 吉林大学社会科学学报. 2010 (4): 131–138.

Hart, Oliver. *Firm, Contract and Financial Structure*. New York: Oxford University Press. 1995.

Huang, Yasheng. "How Unique Is the China Model?—A Comparative Analysis of the Economic Data of China, India and Brazil". *Shenzhen University Journal* (*Humanities & Social Sciences*). 2012 29(1):57–61. 黄亚生. 中国模式到底有多独特—基于中国、印度、巴西经济数据的比较分析. 深圳大学学报(人文社会科学版). 2012 29(1): 57–61.

Huang, Yongfeng and Ruoen Ren. "A Comparative Study of the Manufacturing TFP Between the US and China". *China Economic Quarterly*. 2002, 2(4):

161–180. 黄永峰, 任若恩. 中美两国制造业TFP比较研究. 经济学季刊. 2002, 2(4): 161–180.

Kremer, Michael and Eric Maskin. "Globalization and Inequality". Harvard University, Weatherhead Center for International Affairs, working paper, 2006.

Levchenko, Andrei. "Institutional Quality and International Trade". *Review of Economic Studies*. 2007, 74 (3): 791–819.

Lewis, W. Arthur. "Economic Development with Unlimited Supplies of Labor". *Manchester School of Economic and Social Studies*. 1954, 22: 139–91.

Nie, Huihua. Why are China-made Products Much Cheaper in the US? *The Economists' Saloon*. 2010 (4): 9–13. 聂辉华. 为什么"中国制造"在美国更便宜? 经济学家茶座. 2010 (4): 9–13.

Nie, Huihua and Zhaoyun Zou. "China Should Shift from 'Demographic Dividend' to 'Institutional Dividend'". *International Economic Review*. 2012 (6): 124–135. 聂辉华, 邹肇芸. 中国应从"人口红利"转向"制度红利". 国际经济评论. 2012 (6): 124–135.

Nunn, Nathan. "Relationship-specificity, Incomplete Contracts and Pattern of Trade". *Quarterly Journal of Economics*. 2007, 122 (2): 569–600.

Qin, Xiao. "Turning Institutions and Labor Productivity Dividend". *China Reform (Bimonthly)*. 转向制度和劳动生产率红利. 中国改革. 2011(1): 20–23.

Research Group of Economic Growth and Macroeconomic Stability in China. "The Labor Supply Effect and the Shift of China's Economic Growth Path". Economic Research Journal. 2007: (10): 4–16. 中国经济增长与宏观稳定课题组. 劳动力供给效应与中国经济增长路径转换.经济研究. 2007(10): 4–16.

United Nations. "The World Population Prospects: The 2008 evision". 2009, http://www.n.org/esa/population/unpop.htm.

Wood, Adrian. "Openness and Wage Inequality in Developing Countries: The Latin American Challenge to East Asian Conventional Wisdom". *World Bank Economic Review*. 1997, 11 (1): 33–57.

Wu, Ting. "The Contradiction Between the Ministry of Commerce and the Ministry of Finance over the Last Decade". *Phoenix Weekly*. 2011 (8): 27–29. 吴婷. 商务部与财政部十年抵牾. 凤凰周刊. 2011 (8): 27–29.

Xie, Qianli, et al. Forms of Ownership and the Trend of China's Industrial Productivity. *The Journal of Quantitative & Technical Economic*. 2002, 18(3): 5–17. 谢千里, 罗斯基, 郑玉歆, 王莉. 所有制形式与中国工业的生产率变动趋势. 数量经济技术经济研究. 2002, 18(3): 5–17.

Appendices

Appendix A.1

Proof of Proposition 2.2

According to the objective function and constraint conditions, construct a Lagrangian equation as follows:

$$
\begin{aligned}
L = &(1-\alpha)t + \lambda_1(1-\overline{c}-t) + \lambda_2(k\Delta c + p\rho F_A - \Delta c) \\
&+ \lambda_3(p\rho F_s - k\Delta c) + \lambda_4(1-\underline{c}-t-k\Delta c - F_A) \\
&+ \lambda_5(\alpha t + k\Delta c - F_s)
\end{aligned}
\tag{A.1}
$$

Solving all the first order conditions:

$$\frac{\partial L}{\partial \alpha} = -t + \lambda_5 t = 0 \tag{A.2}$$

$$\frac{\partial L}{\partial t} = 1 - \alpha - \lambda_1 - \lambda_4 + \alpha\lambda_5 = 0 \tag{A.3}$$

$$\frac{\partial L}{\partial F_A} = \lambda_2 p\rho - \lambda_4 = 0 \tag{A.4}$$

H. Nie, *Collusion, Local Governments and Development in China*,
DOI 10.1007/978-981-10-5059-6

$$\frac{\partial L}{\partial F_S} = \lambda_3 p\rho - \lambda_5 = 0 \quad (A.5)$$

$$\lambda_3 = \frac{1}{p\rho} > 0 \quad (A.6)$$

$$\lambda_5 = 1 \quad (A.7)$$

$$\lambda_1 + \lambda_4 = 1 \quad (A.8)$$

$$\lambda_4 = p\rho\lambda_2 \quad (A.9)$$

$$\lambda_5 = p\rho\lambda_3 \quad (A.10)$$

According to Kuhn–Tucker theorem, all Lagrangian multipliers $\lambda_i(\bullet) \geq 0$. When $\lambda_i > 0$, the corresponding constraint conditions take the mark of equality; when $\lambda_i = 0$, the corresponding constraint conditions will be automatically satisfied.

(1) If $\lambda_4 = 0$, then $\lambda_i = 1$, $\lambda_2 = 0$, the constraint conditions are arranged by and as follows:

$$1 - \bar{c} - t = 0 \quad (A.11)$$

$$k\Delta c + p\rho F_A > \Delta c \quad (A.12)$$

$$p\rho F_S = k\Delta c \quad (A.13)$$

$$1 - \underline{c} - t - k\Delta c - F_A > 0 \quad (A.14)$$

$$\alpha t + k\Delta c = F_S \quad (A.15)$$

Then we obtain:

$$t = 1 - \bar{c} \quad (A.16)$$

$$F_A > \frac{(1-k)\Delta c}{p\rho} \quad (A.17)$$

$$F_S = \frac{k\Delta c}{p\rho} \tag{A.18}$$

$$F_A < (1-k)\Delta c \tag{A.19}$$

Obviously, the two inequalities (A.17) and (A.19) are contradictory. Therefore $\lambda_4 > 0$, $\lambda_2 > 0$.

(2) If $\lambda_1 > 0$, the constraint conditions can be arranged by and as follows:

$$1 - \bar{c} - t = 0 \tag{A.20}$$

$$k\Delta c + p\rho F_A = \Delta c \tag{A.21}$$

$$p\rho F_S = k\Delta c \tag{A.22}$$

$$1 - \underline{c} - t - k\Delta c - F_A = 0 \tag{A.23}$$

$$\alpha t + k\Delta c = F_S \tag{A.24}$$

Then we obtain:

$$t = 1 - \bar{c} \tag{A.25}$$

$$F_A = \frac{(1-k)\Delta c}{p\rho} \tag{A.26}$$

$$F_S = \frac{k\Delta c}{p\rho} \tag{A.27}$$

$$F_A = (1-k)\Delta c \tag{A.28}$$

Obviously, the two Eqs. (A.26) and (A.28) are contradictory; therefore $\lambda_1 = 0$, $\lambda_2 = \lambda_3 = \frac{1}{p\rho} > 0$, $\lambda_4 = \lambda_5 = 1$. This implies that the constraint condition (AIR) is loose while others are tight. According to the four tight constraint conditions, the values of all endogenous variables can be figured out. Q.E.D.

Appendix A.2

Proof of Proposition 2.4

When compensation m is added,we get a new Lagrangian equation:

$$\begin{aligned} L =&(1-\alpha)t + \lambda_1(1-\bar{c}-t) + \lambda_2(k\Delta c + p\rho F_A + pm - \Delta c) \\ &+ \lambda_3(p\rho F_S - k\Delta c) + \lambda_4(1-\underline{c}-t-k\Delta c - F_A - m) \\ &+ \lambda_5(\alpha t + k\Delta c - F_S) \end{aligned} \tag{A.29}$$

The first order conditions and Lagrangian multipliers are just the same as those in Appendix A.1.

If $\lambda_4 = 0$, then $\lambda_1 = 1$, $\lambda_2 = 0$. The constraint conditions can be arranged by and as follows:

$$t = 1 - \bar{c} \tag{A.30}$$

$$F_A > \frac{(1-k)\Delta c - pm}{p\rho} \tag{A.31}$$

$$F_A < 1 - \underline{c} - t - k\Delta c - m \tag{A.32}$$

$$F_S = \frac{k\Delta c}{p\rho} \tag{A.33}$$

Slightly modify the form of constraint conditions. Inequality (A.31) is equivalent to

$$p\rho F_A + pm > (1-k)\Delta c$$

Inequality (A.32) is equivalent to

$$F_A + m < 1 - \underline{c} - t - k\Delta c = 1 - \bar{c} - t + (1-k)\Delta c$$

According to the assumption, we know

$$p\rho < 1,\ \ p < 1 \text{ and } 1 - \bar{c} - t \geq 0$$

Inequalities (A.31) and (A.32) cannot be simultaneously satisfied, so the assumption is untenable, and we get $m > \frac{(1-k)(1-p\rho)\Delta c}{p(1-\rho)}$.

(2) If $\lambda_4 > 0$, the following two conditions (a) and (b) should be considered:

a. When $\lambda_1 > 0$, all Lagrangian multipliers are greater than zero, i.e. all constraint conditions are tight. According to the above analysis, it is obvious that (A.31) and (A.32) cannot simultaneously take the mark of equality; therefore, the assumption is untenable, and

$$m = \frac{(1-k)(1-p\rho)\Delta c}{p(1-\rho)}$$

b. When $\lambda_1 = 0$, then all other Lagrangian multipliers are greater than zero, i.e. except (AIR), all constraint conditions are tight.

The solutions are:

$$\alpha^{cp} = \frac{(1-p\rho)k\Delta c}{(1-\underline{c}-m)p\rho - (1-k+p\rho k)\Delta c + pm} \tag{A.34}$$

$$t^{cp} = 1 - \underline{c} - k\Delta c - m - \frac{(1-k)\Delta c - pm}{p\rho} \tag{A.35}$$

$$F_A = \frac{(1-k)\Delta c - pm}{p\rho} \tag{A.36}$$

$$F_S = \frac{k\Delta c}{p\rho} \tag{A.37}$$

Combining the (AIR) condition and Eq. (A.35), $m < \bar{m} = \frac{(1-k)(1-p\rho)\Delta c}{p(1-\rho)}$ is accessible. Intuitively, in the case of information asymmetry, only when the agent gets positive information rent can he be motivated to choose a good mode of production. Therefore, only when compensation is below the critical level $\overline{m}$, can the optimal collusion-proof mechanism be in effect. Then

$\frac{\partial \alpha^{cp}}{\partial m} < 0, \frac{\partial t^{cp}}{\partial m} > 0, \frac{\partial F_A}{\partial m} < 0$. Q.E.D.

Appendix A.3

An Introduction to the Chinese Industrial Enterprise Database[1]

The Chinese Industrial Enterprise Database is created by the China National Bureau of Statistics. The data are mainly from the summaries of quarterly and annual reports presented by the sample enterprises to local bureaus of statistics. The full name of the database is "Database of All State-Owned Industrial Enterprises and Non-State-Owned Industrial Enterprises above Designated Size", and the statistical units are business legal representatives. The statistical scope for "industry" includes the three categories of mining and quarrying, manufacturing, and producers and suppliers of electric power, gas and water based on the Standard Industrial Classification (SIC) Codes, manufacturing being the majority (accounting for 90%) of all the sample industries. "Above designated size" here refers to the prime operating revenue (sales revenue) of the sample enterprises being CNY 5 million and above, which later modified to CNY 20 million and above in 2011. Data based on the above statistical scope began to be collected from 1998, but many experts used the 1999–2007 database. Since the main component of the database is manufacturing enterprises, the statistical scope is basically the same as the industrial classification of other countries, and some invariables (such as capital, R&D inputs and export delivery value) are easier to measure. This is the reason why many users often separate manufacturing industry out of the database. The statistical scope of manufacturing includes thirty categories (2-digit industry) such as agricultural and sideline food processing industry, industrial artifacts, waste recovery and processing, which correspond to the codes 13–43 (excluding 38) Standard Industrial Classification Codes (GB/T4754-2002).

The 1999–2007 Chinese Industrial Enterprise Database has a large collection of more than 2 million observed values. During the sample period, the sample enterprises increased year after year to approximately 330,000 in 2007 from 160,000 in 1999 and it now contains altogether 550,000 enterprises (including listed industrial enterprises).[2] Obviously, there is a huge volume of unbalanced panel data. Owing to bankruptcy, system reform and restructuring and other reasons, only 46,000 enterprises (accounting for about 8%) have always been in the sample period. According to the comparable *China Economic Census Yearbook 2004*, the sales figure of industrial enterprises amounted to CNY 21.84 trillion, while that of the sample enterprises in the Database were CNY19.56 trillion, accounting for 89.5% of the national total.[3] Table A.1 provides information of the changes of the shares of all types of industrial enterprises—state-owned, collective, private, and foreign-capital (including Hong Kong, Macau, and Taiwan) enterprises. The shares of state and collective enterprises decreased significantly from two-thirds in 1999 to less than one-tenth in 2007, while the share of private enterprises increased rapidly from less than 20% in 1999 to more than 70% in 2007. This table reflects, from a profile, the dramatic changes of the Chinese market economy.

The Database is also the most comprehensive enterprise database. It contains two types of information: basic descriptions and financial data. The former includes: legal person code, name of the enterprise, legal representative, telephone number, zip code, detailed address, industrial category, registration type (form of ownership), affiliation, year founded, and number of employees. The latter includes almost 130 indicators: liquid assets, receivables, long-term investment, fixed assets, cumulative depreciation, intangible assets, liquid liabilities, long-term liabilities, paid-in capital, prime operating revenue, prime operating cost, operating expenses, overheads, financial expenses, operating profit, total pre-tax profits, advertising expenses, R&D inputs, gross payroll, employee services and benefits, value-added tax, intermediate inputs, gross value of industrial output, and export delivery value. Especially, the 2004 national economic census includes such indicators as the proportion of male and female employees classified according to their educational backgrounds and their professional and technical titles. It also contains information as to whether a sample enterprise is affiliated to the trade union and how many people join that organization.

Table A.1 Type, Number and Percentage of the Chinese Industrial Enterprises

Year	*State-owned*	*Percentage (%)*	*Collective*	*Percentage*	*Private*	*Percentage*	*Foreign-capital*	*Percentage*	*Total number*
1999	52,817	32.86	53,507	33.29	27,757	17.27	26,652	16.58	160,733
2000	44,665	27.66	49,383	30.58	39,192	24.27	28,240	17.49	161,480
2001	36,781	21.67	42,528	25.06	59,208	34.89	31,178	18.37	169,695
2002	31,570	17.55	38,237	21.25	75,884	42.18	34,208	19.02	179,899
2003	25,157	12.93	32,334	16.62	98,698	50.74	38,318	19.70	194,507
2004	27,403	9.89	26,896	9.70	165,864	59.85	56,976	20.56	277,139
2005	18,520	6.86	23,875	8.84	171,603	63.53	56,112	20.77	270,110
2006	16,209	5.40	20,983	6.99	202,417	67.43	60,585	20.18	300,194
2007	11,724	3.50	19,355	5.78	236,823	70.68	67,174	20.05	335,076

Source Nie et al. (2012)

Appendix A.4

Computational Approach to Total Factor Productivity[4]

The current estimation approaches to total factor productivity are mainly parametric method, nonparametric method and semiparametric method.

Parametric method refers to setting up a production function (often Cobb-Douglass production function) and estimating output elasticity by cost minimization or regression method and then working out Solow residual through ordinary least square (OLS). The advantage of this method is that the simple formula setting-up bears an intuitive economic implication, but the disadvantage is that the setting up of production function may not necessarily corresponds to the reality—for instance, the setting-up of the constant returns to scale and constant elasticity is perhaps too rigid.

Nonparametric method includes data envelopment analysis (and Malmquist index based on it) and stochastic frontier analysis. The advantage of this method is that it directly uses linear programming, without setting a prior transcendental production function, to construct production frontier for observed data and then sets the distance between every enterprise and production frontier as efficiency. It can also decompose total factor productivity into allocative efficiency, scale efficiency, and technological innovation. But its disadvantage is obvious too. Data envelopment analysis ignores the stochasticity of the samples, and stochastic frontier analysis arbitrarily sets up probability distribution of error items. A more serious question is that this method actually assumes that not all enterprises could maximize profits or second best efficiency under the constraint condition.[5]

Semi-parametric method is an OP method which combines production function estimation and nonparametric estimation (Olley and Pakes, 1996). Its advantage is that it can solve the problems of simultaneity and sampling at the same time in estimating TFP,[6] and its disadvantage is about the same as that of parametric method. Due to long inter-temporal sample data which spans over 9 years from 1999 to 2007, high frequency of enterprises' entry into and exit from the database, and significant heterogeneity among enterprises, sampling is really a serious matter. In addition, the impact of the productivity of different enterprises on the continuous factor input cannot be underestimated either. Therefore, it is advisable to choose OP method.

Here is a brief introduction of the fundamentals of the OP method. Without loss of generality, suppose the production function is the Cobb-Douglas form, $Y_{it} = \Omega_{it} L_{it}^{\beta_l} M_{it}^{\beta_m} K_{it}^{\beta_k}$, among which Y stands for output, Ω for TFP, L for labor, M for middleman, K for capital. Taking the natural logarithm on the both sides of the equation, we obtain

$$y_{it} = \beta_0 + \beta_l l_{it} + \beta_m m_{it} + \beta_k k_{it} + \mu_{it} \tag{A.38}$$

According to Solow residual, the constant item and error item in formula (A.38) is TFP. But two problems arise from it: simultaneity and sampling. As to simultaneity, at least part of TFP can be perceived at an earlier date by the enterprise and therefore influence its decision on factor inputs. This implies that the error item is in correlation with explanatory variables, and a simple OLS estimation at this time is biased. As to sampling, those enterprises which still survive in the competitive market are much more productive while those less productive enterprises are likely to be sifted out, so the TFP will usually be overestimated if its estimation is based only on sample enterprises.

For convenience's sake, we might as well decompose the error item into two: $\mu_{it} = w_{it} + \eta_{it}$, among which w denotes TFP, η denotes real error item or productivity shock. Suppose capital accumulation is formed by perpetual inventory method, i.e. $k_{t+1} = (1 - \delta)k_t + i_t$, among which i denotes investment, and δ denotes depreciation. To solve the above two endogenous problems, Olley and Pakes (1996) assumed that investment decision is the function of productivity, capital and intermediate inputs, i.e. $i_t = i_t(w_t, m_t, k_t)$, and substitute it into Eq. (A.38):

$$y_{it} = \beta_l l_{it} + \phi_t(i_{it}, m_{it}, k_{it}) + \eta_{it} \tag{A.39}$$

Equation (A.39) is a semiparametric regression formula which can solve the problem of simultaneity. To solve the problem of sampling, we need to estimate an enterprise survival probability equation:

$$\begin{aligned} \Pr\{\chi_{t+1} = 1 | w_{t+1}(k_{t+1}, m_{t+1}), J_t\} &= \rho_t\{w_{t+1}(k_{t+1}, m_{t+1}), w_t\} \\ &= \rho_t(i_t, m_t, k_t) \equiv P_t \end{aligned} \tag{A.40}$$

J denotes all information in the current period. In the first step, using a four-order poly-nominal to estimate Eq. (A.39), we get β_t and ϕ_t.

In the second step, using probit modeling to estimate Eq. (A.40), we obtain P_t. In the third step, consider the Eqs. (A.39) and (A.40), then minimizing residual sum of squares (RSS), we get β_m and β_k. At last, by Slow residual method, we find:

$$TFP_{it}^{OP} = \exp(y_{it} - \hat{\beta}_l l_{it} - \hat{\beta}_m m_t - \hat{\beta}_k k_{it}) \tag{A.41}$$

The core of the OP method is to count investment as the instrumental variable of productivity. It assumes that there is a strictly monotonical relationship between investment and output, which implies that all the observed values with zero in investment will be eliminated. To solve the problem, Levinson and Petrin (2003) used intermediate input as the proxy variable of productivity in that the observed values with zero in intermediate inputs are fewer than those in investment. However, the LP method fails to solve the problem of sampling, and sometimes some invalid situation may occur when the gradient of the intermediate inputs. In view of high frequency (annual above 10%) of enterprises' entry into and exit from the Chinese Industrial Enterprise Database, we cannot ignore the problem of sampling and therefore choose the OP method over others to estimate total factor productivity.

Notes

1. For a detailed introduction and analysis of this Database, see Nie et al. (2012).
2. This Database is readily available from different sources, but they have little or no differences.
3. Here, the sales figure reported by the economic census is from "Communique on Major Data of the First National Economic Census of China (No. 2)" on the website of the China National Statistical Bureau, while the sales figure of the sample enterprises in the Industrial Enterprise Database is calculated by the author of the book.
4. For specific computational methods and applications, see Nie and Jia (2011).
5. Kumbhakar and Lovell (2003) frankly admit this disadvantage in their classical work *Stochastic Frontier Analysis.*
6. In a specific functional form, stochastic frontier analysis allows for the correlation between inputs and technical efficiency, but it can never solve the problem of sampling.

Further Reading

Bartelsman, Eric J., John C. Haltiwanger, Stefano Scarpetta. "Cross-Country Differences in Productivity: The Role of Allocation and Selection", NBER working paper. No. 15490, 2009.

Bolton, Patric and Mathias Dewatripont. *Contract Theory*. Cambridge: MIT Press, 2005.

Callic, Rowan. "The China Model". *The American*. 2007, 76 (1): 13–16.

Caulfield, Janice. "Local Government Reform in China: A Rational Actor Perspective". International Review of Administrative Sciences. 2006, 72: 253–267.

Chen, Jian and Bo Gao. Income Gap, House Prices and Changes of Expenditure: An Analysis Based on Panel Data Simultaneous Equations Model (1999–2009). Working paper, 2011.

收入差距、房价与消费变动——基于面板数据联立方程模型的分析. 上海经济研究, 2012(2): 53–62.

Jia, Ruixue. "Pollution for Promotion". Stockholm University, IIES working paper, 2012.

Jud, Donald, and Dan Winkler. "The Dynamics of Metropolitan Housing Prices". *Journal of Real Estate Research*. 2002, 23 (1/2): 29–46.

Kahn, Joseph and Jim Yardley. "Choking on Growth: Part I-IX", *The New York Times*, August 26, 2007.

Kofman, Fred and Jacques Lawarree. "Collusion in Hierarchical Agency", *Econometrica*. 1979, 47 (4): 801–819.

Kofman, Fred and Jacques Lawarree. "On the Optimality of Allowing Collusion". Journal of Public Economics. 1996, 61 (3): 383–407.

Kumbahakar, Subal C., and C. A. Knox Lovell. *Stochastic Frontier Analysis*. Cambridge, the UK: Cambridge University Press. 2003.

H. Nie, *Collusion, Local Governments and Development in China*,
DOI 10.1007/978-981-10-5059-6

Lambert-Mogiliansky, Ariane. "On the Optimality of Illegal Collusion in Contracts". Review of Economic Design. 1998, 3: 303–328.

National Bureau of Statistics. *China Statistical Yearbook*. Beijing: China Statistical Publishing House. 1981–2012. 国家统计局. 中国统计年鉴. 北京: 中国统计出版社. 1981–2012.

Nie, Huihua and Qiaochu Li. "A Neo-Political Interpretation of High House Prices in China: From a Perspective of Collusion". *Teaching and Research*. 2013, V47 (1): 50–62. 聂辉华, 李翘楚. 中国高房价的新政治经济学解释——以政企合谋为视角. 教学与研究. 2013, V47 (1): 50–62.

Nie, Huihua, Ting Jiang and Rudai Yang. "The Present Situation of the Use of the Chinese Industrial Enterprise Database and Potential Problems". *The Journal of World Economy*. 2012, 35 (5): 142–158. 聂辉华, 江艇, 杨汝岱. 中国工业企业数据库的使用现状和潜在问题. 世界经济, 2012, 35 (5): 142–158.

Opper Sonja, Stefan Brehm. "Economic Performance and Networks: Political Careers in China's M-Form State". Working paper, 2007.

Ramo, Joshua Cooper. "The Beijing Consensus". London, the Foreign Policy Center, 2004.

Restuccia, Diego and Richard Rogerson. "Policy Distortions and Aggregate Productivity with Heterogeneous Establishments". *Review of Economic Dynamics*. 2008, 11 (4): 707–720.

Shapley, Lloyd. "A Value for n-person Games". in H. W. Kuhn and A. W. Tucker, eds. *Contributions to the Theory of Games*, vol. II. Princeton: Princeton University Press, 1953.

State Administration of Coal Mine Safety. *China Coal Industry Yearbook*. Beijing: Coal Industrial Publishing House. 1982–2012.

Subodh, Kandamuthan. "Market Concentration, Firm Size, Innovative Activity: A Firm-level EconomSic Analysis of Selected Indian Industries under Economic Liberalization". WIDER, Discussion Paper, No. 108, 2002.

Suzuki, Yutaka. "Collusion in Organizations and Management of Conflicts through Job Design and Authority Delegation". Journal of Economic Research. 2007, 12 (2): 203–241.

Svensson, Jacob. "Eight Questions about Corruption". *Journal of Economic Perspectives*. 2005, 19 (3): 19–42.

Syverson, Chad. "What Determines Productivity?" NBER working paper No. 15712, 2010.

Tao, Ran, Fubing Su, Xi Lu and Yuming Zhu. Can Economic Growth Bring about Political Promotion?——A Logical Challenge Against Promotion Tournament Theory and a Provincial-Level Empirical Re-Evaluation. *Management World*. 2012 (12): 13–18. 经济增长能带来竞争吗? ——对晋升锦标赛理论的逻辑挑战与省级实证重估. 管理世界. 2012 (12): 13–18.

Tse, Raymond. *Real Estate Economics: Theory and Policy with Reference to Hong Kong, Singapore and Taiwan*. Hong Kong: EIA Publishing, 1994.

World Bank. *World Development Report 2006: Equity and Development.* A copublication of The World Bank and Oxford University Press. 2005.
Xiao, Xingzhi, Changshi Chen, and Yingfei Qi. "A Research on the Asymmetrical Impact of the Safety Regulations Fluctuations on Coal Production.

Index

H. Nie, *Collusion, Local Governments and Development in China*,
DOI 10.1007/978-981-10-5059-6

The manufacturer's authorised representative in the EU is Springer Nature Customer Service Centre GmbH, Europaplatz 3, 69115 Heidelberg, Germany. If you have any concerns regarding our products, please contact ProductSafety@springernature.com

Printed and bound by CPI Group (UK) Ltd, Croydon, CR0 4YY
25/09/2026
02222897-0009